LANGE

USMLE ROAD MAP

NEUROSCIENCE

Second Edition

JAMES S. WHITE, PhD

Adjunct Assistant Professor of Cell and Developmental Biology
University of Pennsylvania School of Medicine
Philadelphia, Pennsylvania

Assistant Professor of Cell Biology
School of Osteopathic Medicine
University of Medicine and Dentistry of New Jersey
Stratford, New Jersey

 **Medical**

New York   Chicago   San Francisco   Lisbon   London   Madrid   Mexico City
Milan   New Delhi   San Juan   Seoul   Singapore   Sydney   Toronto

4 5 6 7 8 9 0   DOC / DOC   14

ISBN 978-0-07-149623-0
MHID   0-07-149623-8
ISSN 1548-1980

Notice

Medicine is an ever-changing science. As new research and clinical experience broaden our knowledge, changes in treatment and drug therapy are required. The authors and the publisher of this work have checked with sources believed to be reliable in their efforts to provide information that is complete and generally in accord with the standards accepted at the time of publication. However, in view of the possibility of human error or changes in medical sciences, neither the authors nor the publisher nor any other party who has been involved in the preparation or publication of this work warrants that the information contained herein is in every respect accurate or complete, and they disclaim all responsiblity for any errors or omissions or for the results obtained from use of the information contained in this work. Readers are encouraged to confirm the information contained herein with other sources. For example and in particular, readers are advised to check the product information sheet included in the package of each drug they plan to administer to be certain that the information contained in this work is accurate and that changes have not been made in the recommended dose or in the contraindications for administration. This recommendation is of particular importance in connection with new or infrequently used drugs.

This book was set in Adobe Garamond by Pine Tree Composition, Inc.
The editors were Catherine Johnson and Harriet Lebowitz.
The production supervisor was Catherine H. Saggese.
The illustration manager was Armen Ovsepyan.
Graphics and illustrations created by Dragonfly Media Group.
The index was prepared by Pine Tree Composition, Inc.
RR Donnelley was the printer and binder.

This book is printed on acid-free paper.

CONTENTS

USING THE
USMLE ROAD MAP SERIES
FOR SUCCESSFUL REVIEW

What Is the Road Map Series?

Short of having your own personal tutor, the *USMLE Road Map* Series is the best source for efficient review of major concepts and information in the medical sciences.

Why Do You Need A Road Map?

It allows you to navigate quickly and easily through your physiology course notes and textbook and prepares you for USMLE and course examinations.

How Does the Road Map Series Work?

Outline Form: Connects the facts in a conceptual framework so that you understand the ideas and retain the information.

Color and Boldface: Highlights words and phrases that trigger quick retrieval of concepts and facts.

Clear Explanations: Are fine-tuned by years of student interaction. The material is written by authors selected for their excellence in teaching and their experience in preparing students for board examinations.

Illustrations: Provide the vivid impressions that facilitate comprehension and recall.

 Clinical Correlations: Link all topics to their clinical applications, promoting fuller understanding and memory retention.

 Clinical Problems: Give you valuable practice for the clinical vignette-based USMLE questions.

 Matching Problems: Offer quick self-test of basic and clinical facts.

 Explanations of Answers: Are learning tools that allow you to pinpoint your strengths and weaknesses.

To my parents, Maggie and Barclay White; my wife, Kim; and my daughters, Kate and Kristen,

all of whom make whatever I do worthwhile.

ACKNOWLEDGMENTS

I am indebted to Anne Marie Santilli, class of 2010, School of Osteopathic Medicine, UMDNJ, and to J. Douglas Jaffe, D.O., R.N. for their editorial efforts. Special thanks to Chris and Karen Zipp for their contributions to the second edition, and to Hazel Murphy for permission to use images of myelin-stained slides from the Drexel College of Medicine collection.

CHAPTER 1
DEVELOPMENT, GROSS ANATOMY, AND BLOOD SUPPLY OF THE NERVOUS SYSTEM

I. **The nervous system consists of the central nervous system (CNS) and the peripheral nervous system (PNS).**

   A. The **CNS includes the brain in the cranial cavity and the spinal cord** in the vertebral canal (Figure 1–1).

      1. The CNS is organized into **regions of gray matter and white matter**.

      2. Gray matter contains the cell bodies of neurons, their dendrites, and the proximal parts of their axons.

      3. Groups of anatomically or functionally similar neuron cell bodies in the gray matter may be found in a nucleus, a lamina, or a layer.

      4. White matter contains axons of neurons. Groups of anatomically or functionally similar axons may be found in a peduncle, a funiculus, a fasciculus, a lemniscus, or a tract.

      5. Both gray and white matter also contain different types of glial or supporting cells.

   B. The **PNS is composed of spinal nerves, cranial nerves, and autonomic nerves**. The PNS also contains supporting cells (see Figure 1–1).

      1. **Nerves usually contain different combinations of axons of motor neurons and processes of sensory neurons.** Groups of neuronal cell bodies in the PNS are found in sensory or autonomic ganglia.

      2. Thirty-one pairs of **spinal nerves**—8 cervical, 12 thoracic, 5 lumbar, 5 sacral, and 1 coccygeal—enter or exit segmentally from the spinal cord (see Figure 1–1).

         a. Spinal nerves supply structures in the limbs, trunk, and neck; all branches of spinal nerves contain both motor and sensory fibers.

         b. Cutaneous branches of a spinal nerve supply a specific dermatome, the area of skin supplied by the branches of a single spinal nerve.

         c. Muscular branches of a spinal nerve supply a specific myotome, the muscle mass supplied by the branches of a single spinal nerve.

      3. Twelve pairs of **cranial nerves** emerge mainly from the parts of the CNS above the spinal cord (see Figure 1–1) and have Roman numeral designations (I–XII).

         a. Cranial nerves mainly supply structures in the head, in the visceral part of the neck, and in viscera in the thorax and abdomen.

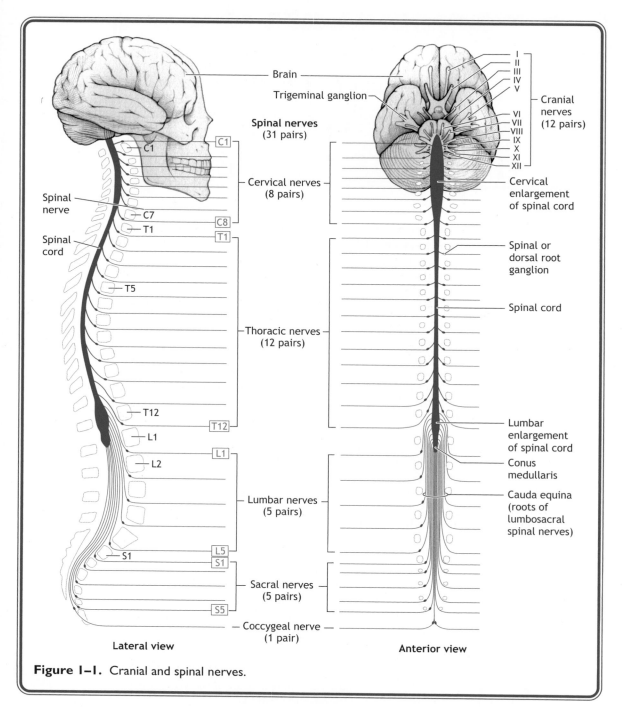

Figure 1-1. Cranial and spinal nerves.

b. Cranial nerves and their branches have motor fibers or sensory fibers or both in different combinations.

    c. At their points of emergence from the CNS, 3 cranial nerves (I, II, and VIII) are considered to be sensory, 5 are considered to be motor (III, IV, VI, XI, and XII), and 4 contain both motor and sensory fibers (V, VII, IX, and X).

4. Autonomic nerves are organized into sympathetic and parasympathetic divisions that provide motor innervation to smooth muscle, glands, and cardiac muscle (Figure 1–2).

    a. The sympathetic and parasympathetic divisions use 2 neurons in series to innervate target structures.

    b. The cell body of the first neuron (the preganglionic neuron) is found in the gray matter of the CNS; the second neuron (the postganglionic neuron) is in an autonomic ganglion in the PNS.

    c. Preganglionic sympathetic neuron cell bodies are found in the thoracic and upper lumbar segments of the spinal cord from T1 through L2 and exit in the ventral roots of spinal nerves from T1 through L2.

    d. Preganglionic parasympathetic cell bodies either are found in the brainstem and exit the brain in 1 of 4 cranial nerves (III, VII, IX, or X) or are found at sacral spinal cord segments S2, 3, and 4 and exit the spinal cord with the ventral roots of S2 through S4 spinal nerves.

    e. The sympathetic division functions as an emergency or catabolic system involved in fight-or-flight responses and has a widespread distribution throughout the body.

    f. The parasympathetic division functions to conserve energy and restore body resources and has a restricted distribution. Parasympathetics supply 2 smooth muscles in the orbit, salivary and mucous glands in the head, and smooth muscle and glands in the walls of thoracic and abdominopelvic viscera.

II. The CNS and PNS develop almost exclusively from ectoderm, beginning in the third week of embryonic life (Figure 1–3A-D).

A. The development of the nervous system begins just after **gastrulation**, which establishes the 3 primary germ layers (ectoderm, mesoderm, and endoderm), which give rise to virtually all the tissues of the body.

    1. The notochord is a rod-like structure that develops in the mesoderm in the embryonic midline.

    2. The notochord produces signaling molecules such as retinoic acid (a derivative of vitamin A) and peptide hormones such as sonic hedgehog that induce the formation of the nervous system.

B. Neurulation begins when the notochord sends signals that induce the overlying ectoderm to form the neural plate (see Figure 1–3C and D).

    1. The **neural plate** marks the initial appearance of the nervous system.

    2. The neural plate is a thickening of ectoderm (now neuroectoderm) in the future dorsal midline of the embryo.

    3. Lateral thickenings of the neural plate form a pair of neural folds separated by the neural groove.

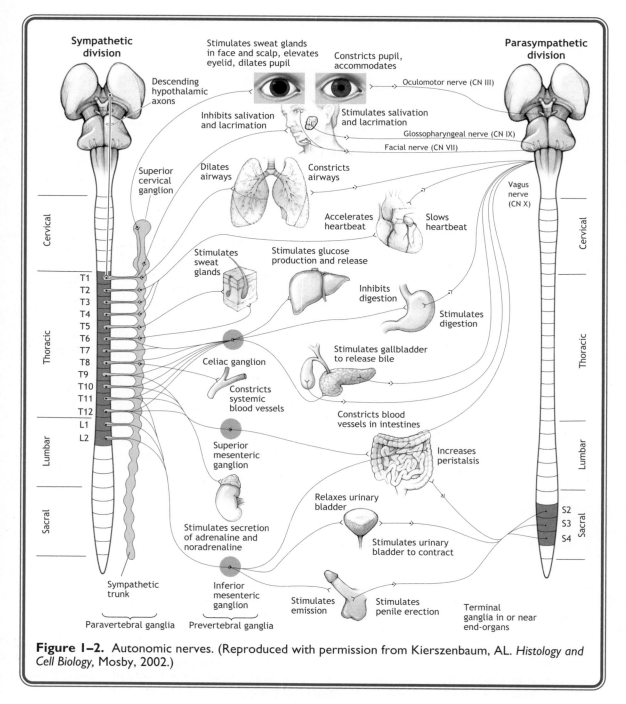

Figure 1–2. Autonomic nerves. (Reproduced with permission from Kierszenbaum, AL. *Histology and Cell Biology,* Mosby, 2002.)

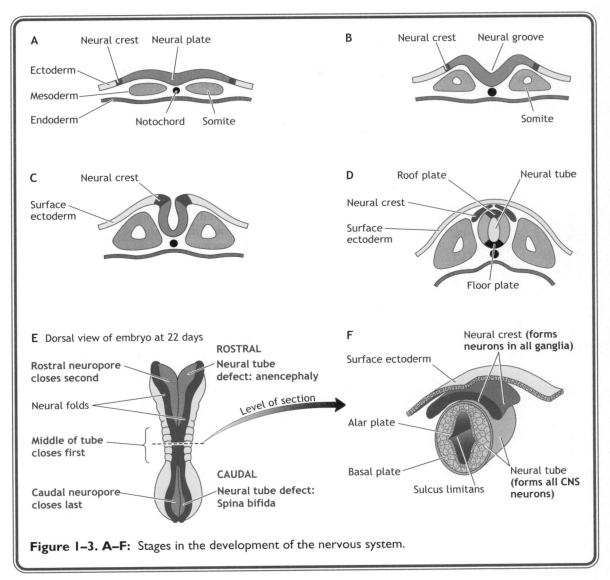

Figure 1–3. A–F: Stages in the development of the nervous system.

4. The **dorsolateral parts of the neural folds** rotate and meet in the dorsal midline, forming the neural tube that surrounds the neural canal.
5. The **neural canal** forms the ventricles of the brain, the cerebral aqueduct, and the central canal.
6. The **middle of the neural tube closes first**, by day 22, leaving open a cranial neuropore and a caudal neuropore; the middle of the neural tube will become the cervical part of the spinal cord (see Figure 1–3E).
7. The **cranial neuropore closes next**, by days 24–25.

8. The **caudal neuropore closes last**, by days 26–27.
9. The neural tube separates from the surface ectoderm, and the surface ectoderm grows over the midline at the site of neurulation.

C. Sonic hedgehog induces the ventral half of the neural tube to form **the floor plate and the basal plate**.

1. The floor plate is a region of glial cells in the ventral midline of the neural tube adjacent to the notochord.
2. The basal plate forms adjacent to the floor plate and gives rise to motor neurons, which innervate skeletal muscle, and preganglionic autonomic neurons.

D. **Bone morphogenetic proteins** (BMPs) induce the dorsal half of the neural tube to form the roof plate and the alar plate.

1. BMPs are expressed by ectoderm cells adjacent to the neural tube.
2. The roof plate is a region of glial cells in the dorsal midline of the neural tube.
3. The alar plate forms adjacent to the roof plate and gives rise to sensory neurons, which respond to inputs provided by sensory roots of spinal and cranial nerves.

E. The **sulcus limitans** is a longitudinal groove that separates the alar plate and the basal plate (see Figure 1–3F).

F. Cells at the dorsolateral margins of the neural plate form **neural crest cells**.

1. Neural crest cells give rise to primary sensory neurons and postganglionic autonomic neurons.
2. The cell bodies of neurons derived from neural crest are found in sensory or autonomic ganglia outside the CNS, except for proprioceptive neurons of the trigeminal nerve, which are found in the mesencephalic trigeminal nucleus.
3. The processes of primary sensory neurons and the axons of postganglionic neurons course in branches of spinal and cranial nerves.
4. Neural crest cells migrate into many parts of the body and form a variety of nonneuronal cell types.

NEURAL CREST CELL MIGRATION FAILURE

Failure of neural crest cells to migrate to specific locations or to differentiate properly may result in a variety of *cardiac, craniofacial, or neurological defects.*

VITAMIN A AND FOLIC ACID LEVELS

Too much or too little vitamin A or a dietary deficiency of folic acid may impede or disrupt closure of the neural tube and cause neural tube defects.

G. **Histogenesis of neurons** begins with rapid mitotic proliferation of cells inside the neural tube.

1. **Neuroblasts** migrate from a ventricular zone to the outer surface of the neural tube using radial glia as a scaffold.
2. **Neuroblast nuclei** form an intermediate zone that develops into gray matter.
3. **Axons that arise from neuroblasts** form a marginal zone that develops into white matter.
4. **Histogenesis** continues in the cerebellum and hippocampus after birth.

FETAL ALCOHOL SYNDROME AND HISTOGENESIS

- *The migration of neuroblasts may be disrupted in fetuses with fetal alcohol syndrome and may result in **microcephaly**, a reduction in the size of gyri in the cerebral hemispheres.*
- *Infants born with microcephaly are prone to be mentally retarded and may develop seizures and motor disorders.*

    H. **Open neural tube defects** most commonly result when the neural folds fail to meet and close the rostral neuropore or the caudal neuropore (see Figure 1–3E).

        1. **Elevated levels of alpha-fetoprotein,** detected by amniocentesis during pregnancy, may indicate the presence of a neural tube defect.

        2. Except for congenital heart defects, **neural tube defects** are the most common serious congenital malformations.

        3. **Abnormalities of the bones of the skull** or vertebrae may also be seen because of a lack of interaction of the neural tube with surrounding mesoderm.

CRANIORACHISCHISIS TOTALIS

- *Craniorachischisis totalis is the most severe open defect.*
- *Craniorachischisis totalis involves a **complete failure of formation of the neural tube**.*

ANENCEPHALY

- *In anencephaly, the cranial neuropore fails to close, and the forebrain fails to develop properly. Anencephaly is not compatible with life; most fetuses succumb in utero or are stillborn.*
- *In an anencephalic fetus, **elevated levels of alpha-fetoprotein** are present, and the pregnancy may be complicated by **polyhydramnios**, an excess of amniotic fluid in the amniotic cavity caused by the absence of the forebrain neurons that control swallowing.*

SPINA BIFIDA

- *Spina bifida results from **defects in the closure of the caudal neuropore or from a failure of the neural tube** to induce the surrounding mesoderm to form the neural arches of 1 or more lumbar or sacral vertebrae.*
- ***Spina bifida cystica** is characterized by a cystlike protrusion through the defect in the neural arches; elevated levels of alpha-fetoprotein are usually evident.*
- *In **spina bifida cystica with meningocele**, the dura and arachnoid line the cyst, and the cyst contains cerebrospinal fluid (CSF) (Figure 1–4A).*
- *In **spina bifida cystica with meningomyelocele**, the most common form of spina bifida cystica, the dura and arachnoid line the cyst, and the spinal cord is displaced into the cyst (see Figure 1–4B).*
- *In spina bifida cystica with meningomyelocele, the lumbosacral spinal nerves may be stretched, and newborn infants may have neurological deficits in the lower limbs.*
- *In **rachischisis**, the caudal neuropore fails to close, the neural arches are absent, and the neural folds are exposed in the lumbosacral region (see Figure 1–4C).*

    I. In a closed neural tube defect, the caudal neuropore closes normally, alpha-fetoprotein levels are normal, and there are no neurological deficits.

        1. In **spina bifida occulta**, the most common closed defect and most common form of spina bifida, there are defects in the neural arches of lumbar or sacral vertebrae; the laminae of the affected vertebrae fail to fuse in the midline to form spinous processes (see Figure 1–4D).

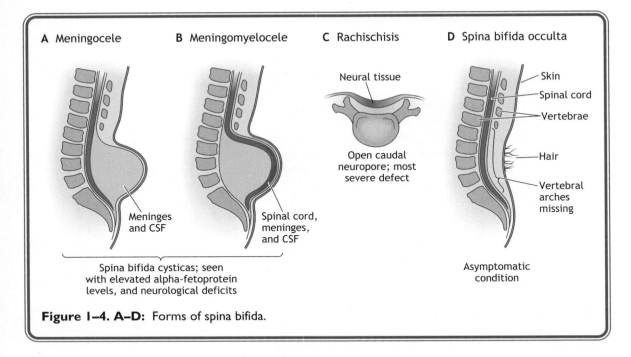

A Meningocele   **B** Meningomyelocele   **C** Rachischisis   **D** Spina bifida occulta

Neural tissue

Open caudal neuropore; most severe defect

Skin
Spinal cord
Vertebrae
Hair
Vertebral arches missing

Meninges and CSF

Spinal cord, meninges, and CSF

Spina bifida cysticas; seen with elevated alpha-fetoprotein levels, and neurological deficits

Asymptomatic condition

Figure 1–4. A–D: Forms of spina bifida.

    2. Often, the only indication of the presence of spina bifida occulta is a tuft of hair in the skin overlying the defect.

III. There are 6 major components of the CNS, 5 of which develop from the 5 secondary vesicles of the neural tube (Figures 1–5 to 1–7).

    A. The **5 secondary vesicles** are the telencephalon, diencephalon, mesencephalon, metencephalon, and myelencephalon.

    B. The **telencephalon** gives rise to the 2 cerebral hemispheres, the preoptic area, and most of the basal ganglia.

       1. Each hemisphere is divided into 4 lobes: frontal, parietal, occipital, and temporal.

       2. Each hemisphere is highly convoluted by gyri that are separated by sulci; the central sulcus and the lateral fissure are deep grooves that partially separate each hemisphere into the frontal, parietal and temporal, and occipital lobes.

    C. The **diencephalon** gives rise to 4 "thalamic" subdivisions: (dorsal) thalamus, hypothalamus, epithalamus, and subthalamus.

       1. The optic nerve [cranial nerve (CN) II] and the optic cup, including the retina, are outgrowths of the diencephalon.

       2. The neurohypophysis, or posterior lobe of the pituitary, is an outgrowth of the hypothalamus.

    D. The **mesencephalon, metencephalon, and myelencephalon** give rise to the midbrain, pons, and medulla, respectively, which form the 3 parts of the brainstem.

       1. The brainstem contains sensory and motor neurons of most of the cranial nerves (III through XII, excluding CN XI, which is a misplaced spinal nerve).

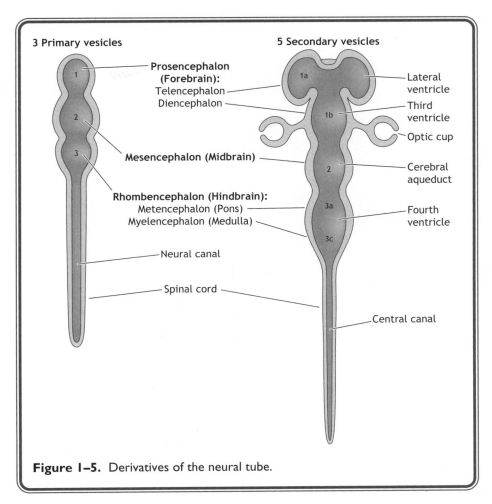

Figure 1–5. Derivatives of the neural tube.

2. The brainstem contains a core of diffusely organized neurons in the reticular formation.

3. The cerebellum is derived from the metencephalon and overlies the pons and medulla.

E. The **spinal cord is the sixth component of the CNS** that extends inferiorly from the medulla but is not derived from a secondary vesicle; the dorsal and ventral roots of 31 pairs of spinal nerves emerge segmentally from the spinal cord.

ENCEPHALOCELE

In an encephalocele, the meninges, the meninges plus part of the CNS, or the meninges, the CNS, and a part of the ventricular system may herniate through a defect in the skull, most commonly in the occipital region.

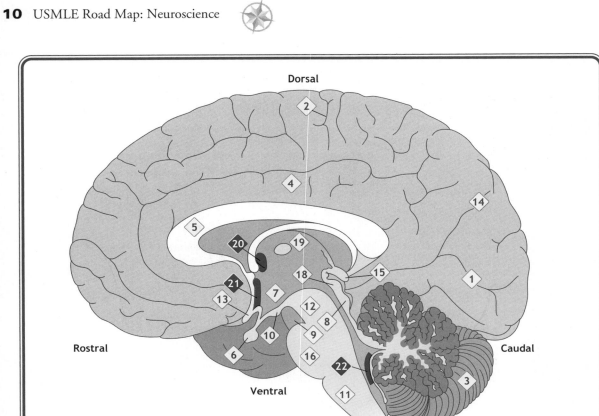

Dorsal

Rostral

Caudal

Ventral

(1) Calcarine sulcus	(7) Hypothalamus	(13) Optic chiasm	(18) Superior colliculus
(2) Central sulcus	(8) Inferior colliculus	(14) Parieto-occipital sulcus	(19) Thalamus
(3) Cerebellar hemisphere	(9) Interpeduncular fossa	(15) Pineal gland	
(4) Cingulate gyrus	(10) Mammillary body	(16) Pons	(20) Subfornical organ
(5) Corpus callosum	(11) Medulla oblongata	(17) Spinal cord	(21) Organum vasculosum
(6) Hypophysis	(12) Midbrain		(22) Area postrema

Figure 1–6. Medial view of the central nervous system.

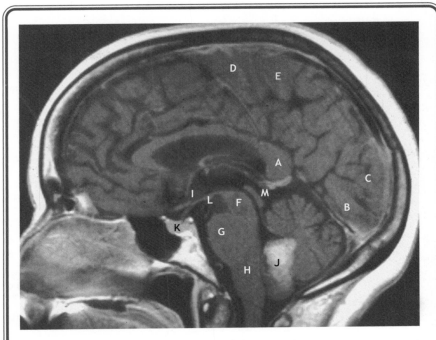

Figure 1–7. MRI of the medial view of the central nervous system.
A. Corpus callosum (splenium). **B:** Lingual gyrus. **C:** Cuneus gyrus. **D:**
Primary motor cortex. **E:** Primary somatosensory cortex. **F:** Midbrain.
G: Pons. **H:** Medulla. **I:** Hypothalamus in wall of third ventricle. **J:** Poste-
rior vermis of cerebellum with intracerebral hemorrhage. **K:** Pituitary.
L: Mammillary body. **M:** Pineal.

ARNOLD-CHIARI MALFORMATIONS

- **Arnold-Chiari I malformation** results from a congenital herniation of the tonsils of the cerebellum inferiorly through the foramen magnum, which may compress the medulla or cervical spinal cord.
- Arnold Chiari II malformation results from a herniation of the cerebellar vermis; type I is seen in young
adults, type II in neonates.
- Arnold-Chiari malformations are commonly seen with a cavitation of the central canal in the spinal
cord (See Figure 3-12) or caudal medulla (a syringomyelia or syringobulbia, respectively). Arnold Chiari
II malformations may also be combined with spina bifida with meningomyelocele.

DANDY-WALKER SYNDROME

In **Dandy-Walker syndrome**, there are congenital defects in the development of the vermis of the cerebellum, and the foramina of Luschka and the foramen of Magendie in the fourth ventricle fail to open.
Patients with Arnold-Chiari malformation and Dandy-Walker syndrome may also have a noncommunicating hydrocephalus.

IV. **The arterial blood supply of most of the CNS is provided by a pair of
internal carotid arteries and a pair of vertebral arteries that give rise to
branches that form the circle of Willis (Figures 1–8 to 1–11).**

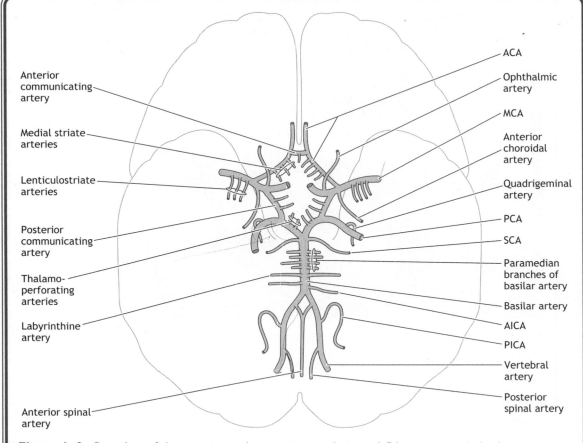

Figure 1–8. Branches of the anterior and posterior circulations. ACA, anterior cerebral artery; MCA, middle cerebral artery; PCA, posterior cerebral artery; SCA, superior cerebellar artery; AICA, anterior inferior cerebellar artery; PICA, posterior inferior cerebellar artery.

A. **Branches of the internal carotid** form the anterior cerebral circulation, and branches of the vertebral arteries form the posterior circulation.

B. At the base of the brain, each internal carotid continues as a middle cerebral artery (MCA) after giving rise to an anterior cerebral artery (ACA); each cerebral artery has superficial and deep branches.

　　1. The MCA passes into the lateral fissure and supplies most of the lateral convexity of each hemisphere (see Figure 1–9).

　　　　a. Two superficial branches of the MCA form a superior division and an inferior division.

　　　　　　(1) The superior division supplies the lateral aspect of the frontal lobe and the anterior part of the parietal lobe.

　　　　　　(2) The inferior division supplies the superior and anterior parts of the temporal lobe and the posterior part of the parietal lobe.

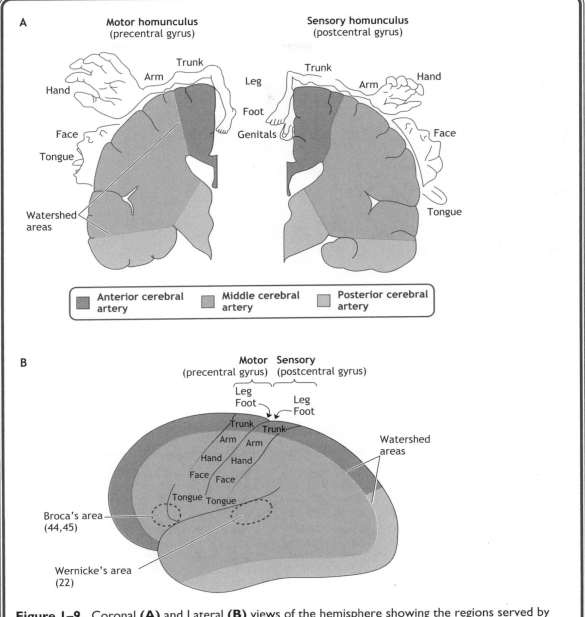

Figure 1–9. Coronal (**A**) and Lateral (**B**) views of the hemisphere showing the regions served by the anterior, middle, and posterior cerebral arteries.

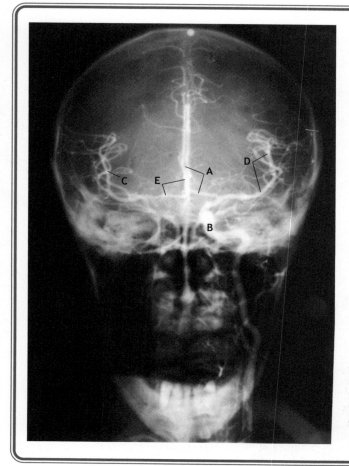

Figure 1–10. Carotid arteriogram (anterior-posterior view). **A:** Left anterior cerebral artery. **B:** Left internal carotid artery. **C:** Right middle cerebral artery. **D:** Left middle cerebral artery. **E:** Right anterior cerebral artery.

    b. The deep branches of the MCA include the anterior choroidal artery, which may branch directly from the internal carotid, and the lenticulostriate arteries (see Figure 1–8).

        (1) The anterior choroidal artery supplies part of the globus pallidus and the optic tract, the genu and posterior limb of the internal capsule, and the hippocampus and amygdala in the medial part of the temporal lobe.

        (2) The lenticulostriate arteries supply most of the putamen and the body of caudate nucleus (parts of basal ganglia).

  2. The ACA is smaller than the MCA and courses through the interhemispheric fissure to the medial aspect of the hemisphere (see Figure 1–8).

    a. The superficial branches of the ACA supply the medial and inferior aspects of the frontal lobe and medial aspect of the parietal lobe (see Figure 1–9).

    b. The deep branches of the ACA are medial striate arteries.

        (1) Occasionally a large medial striate artery, the recurrent artery of Heubner, arises from the proximal part of the ACA.

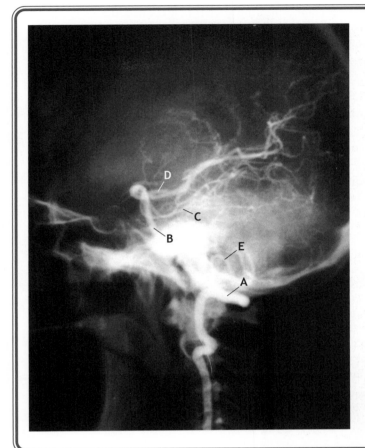

Figure 1–11. Vertebral arteriogram (lateral view). **A:** Vertebral artery. **B:** Basilar artery. **C:** Superior cerebellar artery. **D:** Posterior cerebral arteries (superimposed). **E:** Posterior inferior cerebellar artery.

   (2) The medial striate arteries supply the anterior limb of the internal capsule and the head of the caudate nucleus.
C. The **vertebral arteries** unite to form the basilar artery; the **vertebral and basilar arteries** and their branches form the posterior circulation, which supplies the brainstem and cerebellum (see Chapter 4) and the posterior and inferior parts of each hemisphere (see Figures 1–8 and 1–11).
   1. The basilar artery ends at the rostral end of the pons by bifurcating into a pair of posterior cerebral arteries.
      a. The superficial branches of the posterior cerebral artery (PCA) course posteriorly to supply the occipital lobe and the posterior parts of the parietal and temporal lobes (see Figure 1–9).
      b. The deep branches of the PCA are the thalamoperforating arteries that arise from the proximal part of the PCA.
         (1) The thalamoperforating arteries supply the thalamus, hypothalamus, and subthalamic nucleus (see Figure 1–8).
         (2) Deep branches of the proximal part of the PCA also supply the midbrain.

D. The **circle of Willis** is completed by 3 communicating arteries that connect the anterior and posterior circulations (see Figures 1–8 and 1–10).

   1. A short anterior communicating artery joins the 2 ACAs.

   2. Two posterior communicating arteries each join an internal carotid artery with a PCA.

V. A stroke, or a "brain attack," results in a sudden onset of neurological deficits caused by ischemia or infarction of brain tissue.

   A. A **hemorrhagic stroke** may result in intracerebral or subarachnoid arterial bleeding; 20% of strokes are hemorrhagic strokes.

   1. A hemorrhagic stroke is most commonly caused by a **subarachnoid hemorrhage** (SAH), which results from **blood leaking from a berry aneurysm**.

   a. Berry aneurysms are saccular outgrowths that commonly occur at branch points in the circle of Willis.

   b. Berry aneurysms most commonly occur (85%) in the anterior circulation at branch points of the anterior communicating artery and ACAs (see Figure 1–12), the posterior communicating artery and the internal carotid, or the superior or inferior divisions of the MCA.

   2. A hemorrhagic stroke may also be caused by bleeding from an **arteriovenous malformation** (AVM).

   a. AVMs are a mass of cerebral veins and arteries that are directly interconnected with no intervening capillary beds.

   b. AVMs most commonly occur on the surface of a hemisphere.

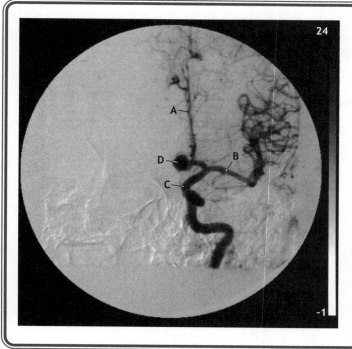

Figure 1–12. Carotid arteriogram with berry aneurysm. **A:** Anterior cerebral artery. **B:** Middle cerebral artery. **C:** Internal carotid artery. **D:** Berry aneurysm at junction of anterior cerebral artery and anterior communicating artery.

BERRY ANEURYSMS AND SUBARACHNOID HEMORRHAGE

- *Blood released from an aneurysm accumulates in the subarachnoid space and irritates the meninges. The **meningeal irritation** causes headaches as well as increased intracranial pressure, vomiting, and an altered level of consciousness.*
- *The optic and oculomotor nerves (CNs II and III) are the nerves most commonly compressed by a berry aneurysm.*

    B. An **ischemic stroke** results from a **vascular occlusion without hemorrhage**; 80% of strokes are ischemic strokes.

    C. Most ischemic strokes involve **large superficial branches of the anterior or posterior circulations** and are commonly caused by a **thrombus** or **embolus** or result from **systemic hypoperfusion**.

        1. A **thrombus** is a clot that forms in a wall of an artery, usually at the site of an **atherosclerotic plaque** that occludes a vessel at that site.

            a. Outside the skull, common sites for the development of a thrombus are at the origin of a vertebral artery from the subclavian or at the origin of an internal carotid from the common carotid.

            b. In the anterior circulation, common sites for the development of a thrombus are in the internal carotid artery proximal to the origins of the MCA and ACA.

            c. In the posterior circulation, common sites for the development of a thrombus are at the site where the vertebral arteries join to form the basilar or at the branch points of the basilar into the PCAs.

THROMBUS

- *A **thrombus in the anterior circulation** may result in a weakness or sensory loss in the contralateral upper limb, lower limb, or face, which may be combined with aphasia, apraxia, or agnosia.*
- *A **thrombus in the posterior circulation** may result in a weakness or sensory loss in the contralateral upper limb, lower limb, or face, which may be combined with cranial nerve signs, gait ataxia, or hemianopsia.*

        2. An **embolus** is a clot that may form at a site of thrombus but dislodges and travels through the bloodstream to occlude a cerebral vessel distal to the thrombus.

            a. Emboli most commonly arise from atheromatous plaques in the extracranial parts of the vertebral or internal carotid arteries.

            b. Cardiac anomalies such as valvular heart disease also commonly give rise to emboli. Air, fat, cholesterol, and protein may also form an embolus.

            c. In the anterior circulation, the superficial branches of the MCA are most commonly occluded by an embolus.

            d. In the posterior circulation, a cerebellar artery or a PCA is most commonly occluded by an embolus.

EMBOLUS

- *An **embolus in the MCA** may result in a weakness or sensory loss in the contralateral upper limb or face, and depending on the hemisphere and branches involved, patients may have aphasia, apraxia, or agnosia.*
- *An **embolus in the posterior circulation** may result in cranial nerve signs, gait ataxia, or hemianopsia.*

3. **Systemic hypoperfusion** may cause an ischemic stroke in watershed or border zones (see Figure 1–9).
   a. **Watershed areas** are regions of overlap at the most distal parts of the vascular territories of the ACA, MCA, and PCA.
   b. The watershed areas are situated on the lateral aspect of each hemisphere between the superficial branches of the MCA and ACA and the superficial branches of the MCA and PCA.
   c. A sudden decrease in systemic arterial pressure may cause ischemia or infarction bilaterally in both the MCA–ACA and the MCA–PCA watershed areas.

WATERSHED INFARCTS AND THE "PERSON IN A BARREL" SYNDROME

- *A **sudden occlusion of an internal carotid artery** may cause unilateral ischemia or infarction in an MCA–ACA watershed area.*
- *These patients may have weakness and sensory deficits in the arm and shoulder, sensory deficits in the trunk (MCA territory), and weakness of thigh muscles and sensory deficits in the thigh (ACA territory).*

D. **Lacunar strokes** are ischemic strokes that involve deep branches of the anterior or posterior circulations.
   1. A **lacunar stroke** most commonly results from a **progressive weakening or narrowing of a deep branch of a cerebral artery** (e.g., medial striate, lenticulostriate, or thalamoperforating arteries) in patients with chronic hypertension or diabetes mellitus.
   2. The structures most commonly affected by lacunar strokes are parts of the basal ganglia or the thalamus, the internal capsule, and the ventral pons or midbrain.

SMALL VESSEL ISCHEMIC STROKES

- *In **lacunar strokes**, patients may have a contralateral hemiparesis, a contralateral hemianesthesia, or a combination of both.*
- ***Aphasias, apraxias, and agnosias** are uncommon in patients with lacunar strokes.*

E. A **transient ischemic attack** (TIA) results from a temporary occlusion of a branch of a cerebral artery.
   1. TIAs cause transient neurological deficits that, by definition, last less than an hour but typically last less than 10 min.
   2. TIAs that persist longer than 10 min may result in an infarction in the brain areas deprived of blood supply and permanent neurological deficits.
   3. A TIA commonly results from small emboli that arise from the site of development of a thrombus and may precede a thrombolytic or embolitic stroke.

TIAS

- *TIAs that result from small emboli shed from a plaque in the internal carotid commonly involve retinal branches of the ophthalmic artery, the first branch of the internal carotid.*
- *Patients may have **amaurosis fugax**, a transient blindness on the affected side that may be perceived as a "window shade" going down over an eye.*

VI. The spinal cord is supplied by direct branches of the aorta and by branches of the vertebral arteries (Figure 1–13).

A. The **cervical and upper thoracic segments of the spinal cord** are supplied by anterior and posterior spinal branches of the vertebral arteries.

B. The **midthoracic segments of the cord** are supplied by individual radicular branches of posterior intercostal arteries that arise from the aorta.

C. The **lower thoracic, lumbar, and sacral segments of the cord** are commonly supplied by a single large radicular artery, the great artery of Adamkiewicz, which arises from a left posterior intercostal artery between T9 and T12.

D. In all regions, the **anterior spinal artery** supplies the white and gray matter in the ventral and lateral parts of the cord.

E. In all regions, a **pair of posterior spinal arteries** supplies the white matter in the dorsal columns and most of the gray matter of the adjacent dorsal horn.

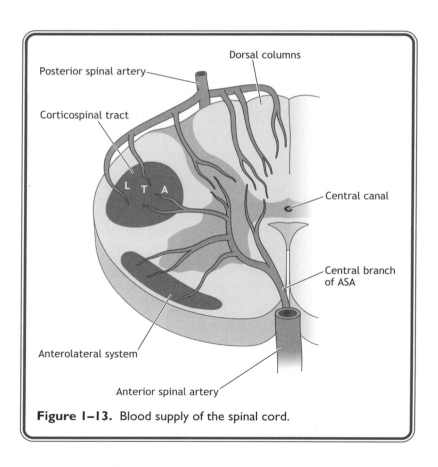

Figure 1–13. Blood supply of the spinal cord.

STROKES INVOLVING ARTERIES THAT SUPPLY THE SPINAL CORD

- *The **anterior spinal vascular territories of thoracic cord segments** are most susceptible to ischemia and infarction as a result of conditions that cause a sudden decrease in aortic pressure.*
- *The **posterior spinal arteries have more potential anastomoses** than the anterior spinal artery, making the anterior spinal artery more dependent on its radicular blood supply.*
- *Patients with a **disruption of the radicular contributors to the anterior spinal artery at thoracic spinal cord segments** have bilateral weakness in the lower limbs, bilateral pain and temperature loss in the lower limbs, and loss of bladder and bowel control (see Chapter 3).*

VII. Cerebral veins drain the brain and empty into a network of dural venous sinuses inside the skull.

    A. Dural venous sinuses are endothelial-lined channels that lack smooth muscle and valves and are situated mainly between the meningeal and periosteal layers of dura.

    B. Most of the **cerebral venous blood** collected in the dural sinuses drains into the internal jugular veins.

    C. Cerebral veins that drain into dural sinuses are known as "bridging veins" because they traverse the subdural space between the arachnoid and the meningeal dura.

    D. The **dural venous sinuses** are sites of resorption of cerebrospinal fluid (CSF); CSF is resorbed from the subarachnoid space through arachnoid granulations that protrude into the sinuses.

SKULL TRAUMA, SUBDURAL AND EPIDURAL HEMATOMAS, AND MASS EFFECT HERNIATIONS

- ***Skull trauma** may cause a shearing of bridging veins at points where they enter dural venous sinuses; venous blood may accumulate in the subdural space and result in a subdural hematoma.*
- *In patients with a **subdural hematoma**, venous blood slowly accumulates in the subdural space and forms a **crescent-shaped hematoma** (Figure 1–14A).*
- *Skull trauma may lacerate a middle meningeal artery and cause a **biconvex, lens-shaped epidural hematoma** adjacent to a hemisphere (Figure 1–14B).*
- *Either type of hematoma may compress part of the brain and cause a **tentorial herniation**, a **tonsillar herniation**, or a **subfalcine herniation**.*
- *In a **tentorial herniation**, the uncus, the most medial part of the temporal lobe, herniates through the tentorial notch of the dura and compresses the brainstem. A PCA may also be compressed by a tentorial herniation.*
- *The patient may have contralateral hemiparesis, a dilated or "blown" pupil resulting from compression of the oculomotor nerve (CN III), and deterioration of cardiovascular and respiratory functions, which may progress to a coma.*
- *In **Kernohan's phenomenon**, the brainstem is compressed against the opposite side of the tentorial notch, resulting in a hemiparesis that is ipsilateral to the herniation. The oculomotor nerve (CN III) on the side of the uncal herniation is compressed in 85% of patients with this condition.*
- *In a **tonsillar herniation**, the tonsil of the cerebellum herniates inferiorly toward the foramen magnum and compresses the medulla, which may result in respiratory compromise and death.*
- *In a **subfalcine herniation**, the cingulate gyrus herniates medially under the falx cerebri, the fold of meningeal dura in the interhemispheric sulcus, and may compress an ACA.*

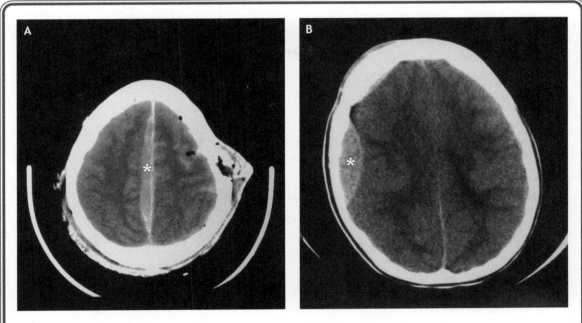

Figure 1–14. Imaging views of a subdural hematoma **(A)** (asterisk) and an epidural hematoma **(B)** (asterisk).

VIII. Circumventricular organs (CVOs) of the brain act as "receptors" that influence hormone secretion or secrete hormones.

   A. CVOs are encompassed by dense beds of fenestrated capillaries that lack the typical features of capillaries at the blood-brain barrier (see Chapter 2).

   B. The **area postrema, the subfornical organ, and the organum vasculosum** are CVOs that act as receptors (see Figure 1–6).

      1. The area postrema is in the wall of the fourth ventricle and responds to circulating toxins that may cause vomiting.

      2. The subfornical organ is in the wall of the third ventricle and responds to circulating levels of angiotensin II.

      3. The organum vasculosum is in the wall of the third ventricle and responds to changes in plasma osmolarity.

      4. Both the subfornical organ and the organum vasculosum regulate the secretion of vasopressin by the hypothalamus.

   C. The **median eminence** and the **pineal gland** are CVOs that are secretory (see Figure 1–6).

      1. The median eminence is adjacent to the infundibulum on the floor of the third ventricle and contains axons that transmit vasopressin and oxytocin from the hypothalamus to the neurohypophysis.

2. The pineal gland is situated posterior to the third ventricle and secretes melatonin, serotonin, and cholecystokinin in response to the circadian rhythm.

IX. The ventricular system functions to protect the brain from trauma and transport CSF.

A. The **brain and spinal cord float within a protective bath of CSF**, which is secreted continuously by the choroid plexus into the ventricles inside the CNS and circulates through the ventricles and through the subarachnoid space (Figures 1–15 and 1–16).

B. **Each part of the CNS contains a ventricle or a channel that interconnects ventricles**; there are 2 lateral ventricles, a third and fourth ventricle (see Figure 1–15).

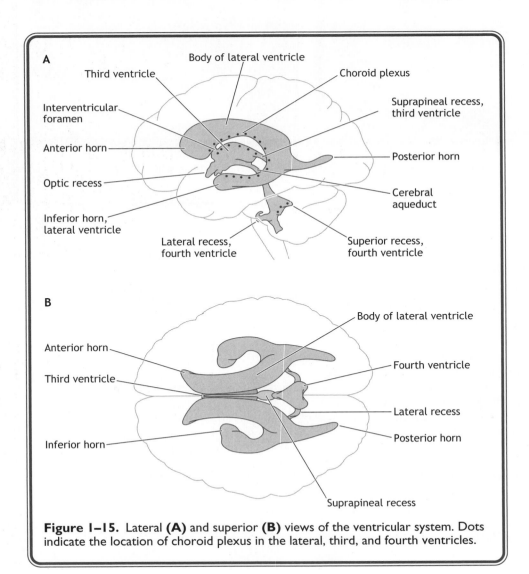

Figure I–I5. Lateral **(A)** and superior **(B)** views of the ventricular system. Dots indicate the location of choroid plexus in the lateral, third, and fourth ventricles.

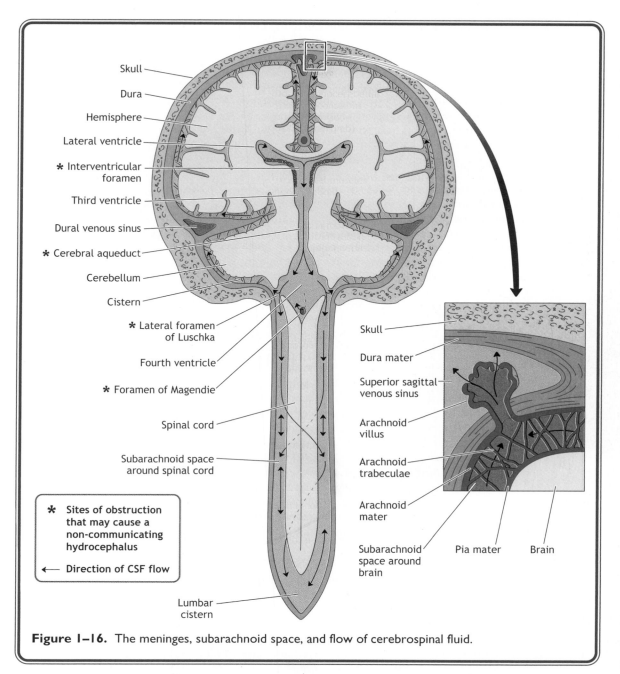

Skull

Dura

Hemisphere

Lateral ventricle

* Interventricular
  foramen

Third ventricle

Dural venous sinus

* Cerebral aqueduct

Cerebellum

Cistern

* Lateral foramen
  of Luschka

Fourth ventricle

* Foramen of Magendie

Spinal cord

Subarachnoid space
around spinal cord

* Sites of obstruction
  that may cause a
  non-communicating
  hydrocephalus

← Direction of CSF flow

Skull

Dura mater

Superior sagittal
venous sinus

Arachnoid
villus

Arachnoid
trabeculae

Arachnoid
mater

Subarachnoid
space around
brain

Pia mater

Brain

Lumbar
cistern

Figure 1–16. The meninges, subarachnoid space, and flow of cerebrospinal fluid.

1. Each lateral ventricle is located within a cerebral hemisphere.
   a. The **lateral ventricles are C shaped** and follow, roughly, the form of the cerebral hemispheres. They have parts situated in each of the 4 anatomic lobes.
   b. The body of each lateral ventricle is in the parietal lobe, the anterior horn extends into the frontal lobe, the inferior horn is in the temporal lobe, and the posterior horn is in the occipital lobe.
   c. Each lateral ventricle communicates with the third ventricle by way of an interventricular foramen (of Monro); the interventricular foramen is at the junction of the anterior horn and the body of each lateral ventricle.
2. The third ventricle is found in the midline between parts of the diencephalon and communicates with the fourth ventricle by way of the cerebral aqueduct (of Sylvius), which passes through the midbrain.
3. The fourth ventricle is located between the dorsal surfaces of the pons and upper medulla and the ventral surface of the cerebellum.
4. The fourth ventricle is continuous with the central canal in the lower medulla, which extends through the length of the spinal cord.
5. The fourth ventricle contains the only openings where CSF can exit the ventricular system and enter the subarachnoid space; there are 2 lateral foramina of Luschka and a single foramen of Magendie in the midline.

C. About 70% of the CSF is secreted by choroid plexus, which consist of glomerular tufts of capillaries covered by choroid epithelial cells (the remaining 30% of CSF represents metabolic waste discharged by the CNS into the ventricles).

1. Choroid plexus are located in the body and inferior horn of each lateral ventricle, in the third ventricle, and in the fourth ventricle (see Figure 1–15A).
2. CSF that is secreted into each lateral ventricle passes through an interventricular foramen into the third ventricle, through the cerebral aqueduct, and into the fourth ventricle.
3. CSF exits the fourth ventricle through the 2 foramina of Luschka and the foramen of Magendie and circulates in the subarachnoid space over the convexity of the brain and around the spinal cord.
4. Most CSF is returned to the venous system by way of arachnoid granulations that protrude from the subarachnoid space into the superior sagittal dural venous sinus (see Figure 1–16).

D. The **total volume of CSF turns over several times a day**.

1. The adult ventricular system and subarachnoid space contain 90–150 mL of CSF, whereas 400–500 mL are produced daily. Only 25 mL of CSF is found in the ventricles themselves.
2. Normal CSF pressure ranges from 60–170 mm H_2O.
3. Normal CSF is a clear fluid that is isotonic with serum (290–295 mOsm/L).
4. The pH of CSF is 7.33, slightly less than that of arterial blood (pH 7.40).
5. CSF has a slightly lower concentration of glucose than serum (60 mg/dL vs. 90 mg/dL) and a significantly lower concentration of protein than serum (35 mg/dL vs. 7000 mg/dL).
6. Normal CSF contains very few white blood cells ($< 4/mm^3$) and no red blood cells.

HYDROCEPHALUS

- *Hydrocephalus* results from an increase in the volume of CSF, which may result in dilatation of 1 or more ventricles.
- A *communicating hydrocephalus* commonly results from impaired absorption of CSF at the arachnoid granulations or a tumor in the subarachnoid space that impedes CSF flow. In these patients, there is increased volume of CSF in the ventricles and in the subarachnoid space.
- A *normal pressure hydrocephalus* is a form of a communicating hydrocephalus that is common in the elderly and is also caused by impaired absorption.
- These patients may have chronically enlarged ventricles and suffer mental decline, urinary incontinence, and an abnormal gait (i.e., "wacky," "wet," and "wobbly"). The abnormal gait is an apraxic gait; there is no weakness, but the patient shuffles as if the feet are stuck to the floor.
- A *noncommunicating hydrocephalus* results from an obstruction to CSF flow inside the ventricular system. The most common site of an obstruction that causes a noncommunicating hydrocephalus in infants is in the cerebral aqueduct. Other sites of obstruction may be at an interventricular foramen or in the fourth ventricle at a foramen of Luschka or the foramen of Magendie (see Figure 1–16).
- A *hydrocephalus ex vacuo* refers to an increase in ventricular size and an increase in the volume of CSF secondary to a pathological loss of brain tissue.

Changes in CSF and Diseases of the CNS

- *Change in the chemical composition of CSF* has diagnostic value in patients with CNS disease.
- In multiple sclerosis (MS) and other inflammatory CNS diseases, the immunoglobulin protein content of CSF increases dramatically.
- *Oligoclonal immunoglobulin bands* are detected by electrophoresis in 90% of patients with MS.
- A *decrease in the glucose concentration in CSF* is seen in patients with **acute bacterial infections** and in those with **meningeal tumors**.
- In patients with **bacterial meningitis**, white blood cell counts are elevated in CSF and may exceed 4000/mm^3.
- Red blood cells may be present in CSF as a result of a **subarachnoid hemorrhage** or a **bloody lumbar puncture**.

Causes, Signs, and Symptoms of Elevated Intracranial Pressure

- A **space-occupying lesion in the skull or vertebral canal** may cause an increase in intracranial CSF pressure. These lesions may raise CSF pressure to 200–600 mm H_2O.
- Space-occupying lesions include tumors, hematomas, or hemorrhages that result from vascular disease or trauma, abscesses, and AVMs.
- **Signs of elevated intracranial pressure** include headache, mental status changes with altered level of consciousness, papilledema, and, less frequently, projectile vomiting.
- Elevated intracranial pressure may also result in **Cushing's triad** of hypertension, bradycardia, and irregular respirations.

CLINICAL PROBLEMS

1. A tumor is growing in the third ventricle. Which of the following parts of the CNS might be compressed?

    A. The mesencephalon

    B. The hypothalamus

    C. The adenohypophysis

D. The hippocampus

   E. The superior colliculi

2. Excessive use of alcohol during a pregnancy causes fetal alcohol syndrome, which affects the development of neural crest cells. Which of the following might be evident in the newborn infant?

   A. There may be fewer neuron cell bodies in the gray matter of the CNS.

   B. The sulci that separate cortical gyri may be wider than normal.

   C. The fetus may have anencephaly.

   D. The optic nerves may be absent.

   E. There may be fewer postganglionic autonomic neurons.

3. Imaging reveals that your patient has a form of hydrocephalus that has resulted in an enlarged lateral ventricle in the right hemisphere. All of the other ventricles appear to be of normal size. Where might there be an obstruction?

   A. At an arachnoid granulation

   B. In the third ventricle

   C. At the foramen of Magendie

   D. At the foramen of Monro on the right

   E. At a foramen of Luschka on the right

4. In this same patient, what form of hydrocephalus is present?

   A. A normal pressure hydrocephalus

   B. A noncommunicating hydrocephalus

   C. A communicating hydrocephalus

   D. Hydrocephalus ex vacuo

5. A young male child is born with a cystlike protrusion in the dorsal midline of the back at the level of the lower lumbar spine. Imaging and diagnostic testing reveal that the cyst contains CSF, and the spinal cord is displaced into the cyst. You diagnose the condition as:

   A. Spina bifida with meningomyelocele

   B. Spina bifida occulta

   C. Rachischisis

   D. Spina bifida with meningocele

   E. Anencephaly

6. An infant is born with a congenital herniation of the tonsils and vermis of the cerebellum inferiorly through the foramen magnum. What else might be evident in the patient?

   A. Syringomyelia

   B. Hirschsprung's disease

   C. Anencephaly

D. Absence of the basal plate

    E. Spina bifida occulta

7. The telencephalon has failed to develop properly. Which part of the brain may be adversely affected?

    A. The mammillary bodies

    B. The hippocampus

    C. The neurohypophysis

    D. The thalamus

    E. The pineal gland

8. An evaluation of a fetus during pregnancy reveals elevated levels of alpha-fetoprotein during amniocentesis, polyhydramnios, and a neural tube defect. Which of the following defects might the fetus have?

    A. Spina bifida with meningocele

    B. Dandy-Walker syndrome

    C. Anencephaly

    D. Rachischisis

    E. Arnold-Chiari malformation

9. An ultrasound image of a fetus reveals the presence of spina bifida occulta. Which of the following statements might apply to this congenital defect?

    A. At birth there may be neurological deficits in the lower limbs.

    B. Amniocentesis during pregnancy may reveal higher than expected levels of alpha-fetoprotein.

    C. The defect is caused by failure of the rostral end of the neural tube to close.

    D. Polyhydramnios may be evident during the pregnancy.

    E. The neural tube has failed to induce the formation of the neural arch at lumbosacral vertebral levels.

10. Your patient develops transient blindness in the right eye that clears after several minutes. What was the most likely cause?

    A. A berry aneurysm compressing the optic chiasm

    B. Emboli in the central artery of the retina on the right

    C. A lacunar stroke involving the right posterior cerebral artery

    D. A watershed infarct in the right MCA–PCA territory

    E. A thrombus in the basilar artery

11. Your patient has suffered from a small lacunar infarct of the thalamus. Occlusion of which of the following vessels might cause the infarct?

    A. Medial striate artery

    B. Deep branch of the PCA

    C. Recurrent artery of Heubner

D. Lenticulostriate artery

   E. Anterior choroidal artery

12. A 49-year-old male patient is brought to the emergency room by his wife because he has difficulty raising his right arm. He is able to make a fist with his right hand and hold objects with no weakness. He complains of numbness on the right side of his trunk and thigh, and he has some weakness of the anterior thigh muscles. A carotid bruit indicates that the left internal carotid artery is stenotic. The patient's signs and symptoms may have resulted from:

   A. Bleeding from an AVM

   B. A subarachnoid hemorrhage

   C. A watershed infarct in the MCA–ACA territory

   D. A lacunar stroke

   E. An embolus in the MCA

13. A patient vomits violently in response to toxins present in CSF. What area of the CNS responded to the toxins and initiated the reflex vomiting?

   A. Organum vasculosum

   B. Area postrema

   C. Subfornical organ

   D. Median eminence

   E. Pineal gland

14. A patient develops a choroid plexus papilloma that results in an oversecretion of CSF. Which of the following might be a location of the papilloma?

   A. Posterior horn of the lateral ventricle

   B. Cerebral aqueduct

   C. Central canal

   D. Anterior horn of the lateral ventricle

   E. Inferior horn of the lateral ventricle

15. A patient has valvular heart disease that gives rise to an embolus, which enters an internal carotid artery. In which branch of the anterior circulation is the embolus most likely to become lodged?

   A. The anterior communicating artery

   B. An MCA

   C. A posterior communicating artery

   D. An ACA

   E. A vertebral artery

16. A 22-year-old male college student suffers head trauma after being thrown from a motorcycle and briefly loses consciousness. A neurological exam given in the emergency room is normal, and the patient is discharged. Several hours later, he is taken

back to the emergency room because he has become drowsy and confused. His right pupil is 5 mm, but his left pupil is only 3 mm in diameter; both react to light. Muscle stretch reflexes are elevated in the left lower limb, and on his return to the emergency room, the patient's left lower limb seemed weaker than the right. A computed tomography scan reveals intracranial bleeding that forms a lens-shaped hematoma between the skull and the lateral aspect of the right hemisphere and subsequent herniation of brain tissue. How would you characterize the intracranial bleeding?

   A. Subdural hematoma

   B. Rupture of an AVM

   C. Epidural hematoma

   D. Subarachnoid hemorrhage

   E. Intraventricular hematoma

17. How would you characterize the herniation in this patient?

   A. Subfalcine

   B. Arnold-Chiari

   C. Tentorial

   D. Tonsillar

   E. Kernohan's

MATCHING PROBLEMS

Questions 18–27: From the following choices, match the embryological division from which the structures are derived to the appropriate structure listed below.

Choices (each choice may be used once, more than once, or not at all):

   A. Mesencephalon

   B. Myelencephalon

   C. Telencephalon

   D. Metencephalon

   E. Diencephalon

   F. Spinal cord

   G. Neural crest

   H. None of the above

18. Neurohypophysis

19. Striatum

20. Retina

21. Cerebellum

22. Hypothalamus

23. Preganglionic sympathetic neurons

24. Pineal gland

25. Inferior colliculus

26. Adenohypophysis

27. Dorsal root ganglia

Questions 28–33: Match the condition in A–F with the most appropriate description.

Choices (each choice may be used once, more than once, or not at all):

    A. Meningocele

    B. Meningomyelocele

    C. Anencephaly

    D. Spina bifida occulta

    E. Rachischisis or myeloschisis

    F. More than one choice in A–E is correct

28. Newborn has a dorsal midline cyst containing CSF.

29. Newborn has a tuft of hair growing in skin over missing spinous processes.

30. Elevated levels of alpha-fetoprotein were evident in amniocentesis during pregnancy; amniotic fluid levels were normal.

31. Normal appearing spinal cord is displaced into a dorsal midline cyst.

32. Caudal neuropore fails to close.

33. Elevated levels of alpha-fetoprotein and polyhydramnios were evident in amniocentesis during pregnancy.

Questions 34–38: Hydrocephalus match

Choices (each choice may be used once, more than once, or not at all):

    A. Noncommunicating hydrocephalus

    B. Communicating hydrocephalus

    C. Hydrocephalus ex vacuo

    D. Normal pressure hydrocephalus

    E. More than one choice in A–D is correct

34. Patient has a tumor that is compressing the cerebral aqueduct.

35. Patient has a tumor in the subarachnoid space.

36. Patient has enlarged ventricles secondary to a loss of neurons due to Alzheimer's disease.

37. Patient has CSF absorption problems at the level of the arachnoid granulations.

38. Patient has bladder problems and walks as if his/her feet are stuck to the floor.

ANSWERS

1. The answer is B. The hypothalamus, which is part of the diencephalon, is situated adjacent to the third ventricle. The colliculi are part of the mesencephalon adjacent to the cerebral aqueduct. The adenohypophysis is not situated near a ventricle, and the hippocampus is situated adjacent to the inferior horn of the lateral ventricle.

2. The answer is E. Of the choices given, the postganglionic autonomic neurons are the only neurons that are derived from neural crest cells. Gray matter and the optic nerve are derived from the neural tube, and anencephaly is a neural tube defect not directly associated with neural crest cells.

3. The answer is D. Based on the direction of CSF flow, an enlarged ventricle should be proximal to an obstruction. An obstruction at the foramen of Magendie or a foramen of Luschka may result in an enlargement of all ventricles; an obstruction in the third ventricle might enlarge both lateral ventricles. An obstruction at an arachnoid granulation may not result in any ventricular enlargement.

4. The answer is B. A noncommunicating hydrocephalus results from an obstruction to CSF flow inside the ventricles or the channels that interconnect them.

5. The answer is A. In spina bifida cystica with meningomyelocele, the most common form of spina bifida cystica, the dura and arachnoid line the cyst, and the spinal cord is displaced into the cyst.

6. The answer is A. The infant has an Arnold-Chiari malformation, which results from a congenital herniation of the tonsils and vermis of the cerebellum inferiorly through the foramen magnum. The Arnold-Chiari malformation is commonly seen with a cavitation of the central canal in the spinal cord or caudal medulla (a syringomyelia) and with spina bifida with meningomyelocele.

7. The answer is B. All other choices are derived from the diencephalon.

8. The answer is C. Spina bifida with meningocele and rachischisis are caudal neural tube defects that may be associated with elevated levels of alpha-fetoprotein during pregnancy, but polyhydramnios is evident only in the event of the rostral neural tube defect. Dandy-Walker syndrome and Arnold-Chiari malformation are caudal defects that would not be causes of polyhydramnios.

9. The answer is E. In spina bifida occulta, the caudal neuropore closes normally, alpha-fetoprotein and amniotic levels are normal, and there are no neurological deficits.

10. The answer is B. TIAs that result from small emboli shed from a plaque in the internal carotid commonly involve retinal branches of the ophthalmic artery, the first branch of the internal carotid.

11. The answer is B. All of the other choices may be involved in a lacunar stroke but not specifically to the thalamus.

12. The answer is C. The patient has "person in a barrel" syndrome with involvement of cortical areas supplied by both the ACA and the MCA.

13. The answer is B. The area postrema is a circumventricular organ in the wall of the fourth ventricle and responds to circulating toxins by inducing vomiting.

14. The answer is E. None of the other choices contain choroid plexus.

15. The answer is B. The MCA is the larger of the 2 branches of the anterior circulation. The PCA and vertebral artery are part of the posterior circulation, and the posterior communicating artery links the 2 cerebral circulations.

16. The answer is C. Skull trauma has lacerated a middle meningeal artery and caused a biconvex, lens-shaped epidural hematoma adjacent to a hemisphere.

17. The answer is C. The epidural hematoma has compressed the lateral part of the right hemisphere and caused a tentorial herniation. In a tentorial herniation, the uncus, the most medial part of the temporal lobe, herniates through the tentorial notch of the dura and compresses the midbrain and the oculomotor nerve.

18. E

19. C

20. E

21. D

22. E

23. F

24. E

25. A

26. H (Oral ectoderm of Rathke's Pouch)

27. G

28. F (both A and B apply)

29. D

30. F (both A, B, and E apply)

31. B

32. E

33. C

34. A

35. B

36. C

37. E (both B and D apply)

38. D

CHAPTER 2
CYTOLOGY OF THE NERVOUS SYSTEM

I. Neurons

A. Neurons are the **functional units of the central nervous system (CNS) and the peripheral nervous system (PNS).**

B. **Neurons are morphologically and functionally polarized** so that information may pass from one end of the cell to the other.

C. Neurons may be classified by the form and number of their processes as **bipolar, unipolar, or multipolar.**

D. The cell **body of the neuron contains the nucleus and membrane-bound cytoplasmic organelles** typical of a eukaryotic cell, including **endoplasmic reticulum, Golgi apparatus, mitochondria, and lysosomes** (Figures 2–1 and 2–2).

   1. The **nucleus and nucleolus** are prominent in neurons.
   2. The **cytoplasm** contains Nissl substance, clumps of rough endoplasmic reticulum with bound polysomes.
   3. The cytoplasm also contains free polysomes; free and bound polysomes in the Nissl substance are sites of protein synthesis.

CNS DISEASE AND CYTOPLASMIC INCLUSIONS IN NEURONS

CLINICAL CORRELATION

- **Lewy bodies** are eosinophilic cytoplasmic inclusions of degenerating neurons in the substantia nigra, pars compacta, in patients with Parkinson's disease, and in cortical and brainstem neurons in patients with certain forms of dementia.
- **Negri bodies** are eosinophilic cytoplasmic inclusions seen in degenerating neurons in the hippocampus and cerebellar cortex in patients with rabies.

E. The **cytoskeleton of the neuron** consists of microfilaments, neurofilaments, and microtubules (see Figure 2–2).

   1. **Neurofilaments** provide structural support for the neuron and are most numerous in the axon and the proximal parts of dendrites.
   2. **Microfilaments** form a matrix near the periphery of the neuron and consist of polymers of actin.
      a. A microfilament matrix is prominent in growth cones of neuronal processes and functions to aid in the motility of growth cones during development.
      b. A microfilament matrix is prominent in dendrites and forms structural specializations at synaptic membranes.

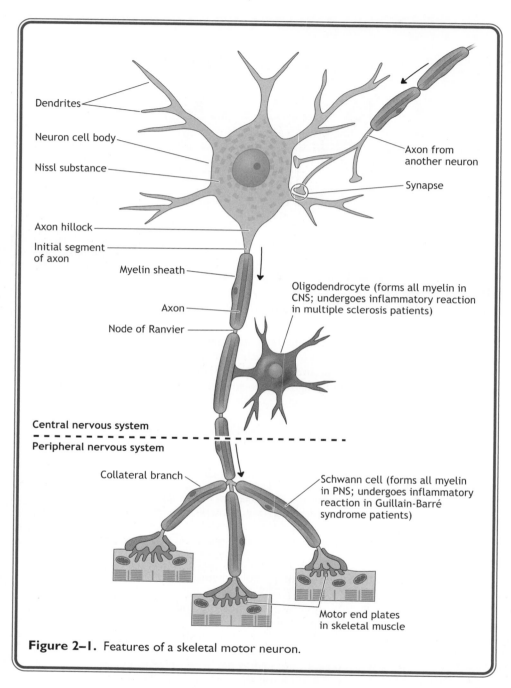

Figure 2–1. Features of a skeletal motor neuron.

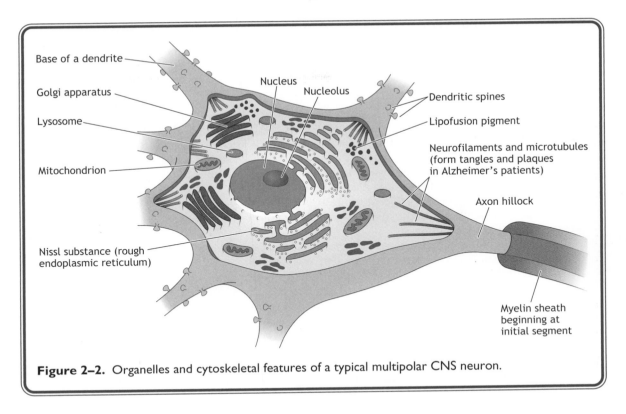

Base of a dendrite

Golgi apparatus

Lysosome

Nucleus

Nucleolus

Dendritic spines

Lipofusion pigment

Neurofilaments and microtubules (form tangles and plaques in Alzheimer's patients)

Mitochondrion

Axon hillock

Nissl substance (rough endoplasmic reticulum)

Myelin sheath beginning at initial segment

Figure 2–2. Organelles and cytoskeletal features of a typical multipolar CNS neuron.

3. **Microtubules** consist of arrays of alpha- and beta-tubulin subunits, are found in all parts of the neuron, and are the cytoplasmic organelles used in axonal transport.

MICROTUBULES AND NEURONAL DEGENERATIVE DISEASES

- In certain degenerative neuronal diseases of the CNS, a **tau protein** becomes excessively phosphory-lated, which prevents this protein from cross-linking microtubules. The affected microtubules form helical filaments, which progress to neurofibrillary tangles and senile plaques in the cell body and dendrites of neurons.
- **Neurofibrillary tangles** are prominent features of degenerating neurons in Alzheimer's disease, amyotrophic lateral sclerosis, and Down syndrome patients.
- In **Alzheimer's disease,** neurons in the limbic system, cholinergic neurons in the basal nucleus of Meynert, and neurons in the cerebral cortex are preferentially affected.

F. **Dendrites** are tapered extensions of the cell body and provide the major surface for synaptic contacts with axons of other neurons (see Figures 2–1 and 2–2).
   1. **Dendrites may contain spines,** which are small cytoplasmic extensions that dramatically increase the surface area of dendrites.
   2. Dendritic **spines contain actin microfilaments** and are motile.
   3. Dendrites may be highly branched; the branching pattern of dendrites may be used to define a particular neuronal cell type.

LOSS OF DENDRITIC SPINES

- *A progressive **loss of dendritic spines** occurs as part of aging, and a severe loss of spines is associated with cognitive disorders.*
- *There is a significant absence of dendritic spines in many types of neurons in patients with **Down syndrome.***

    G. The **axon** (1 per neuron) is specialized for conduction of neural electrical activity away from the cell body to its targets (Figures 2–1, 2–2, and 2–3).

        1. An axon has a **uniform diameter and may branch at right angles into collaterals** along the length of the axon, in particular near its distal end.

        2. The proximal part of the axon is usually marked by an **axon hillock,** a cone-like extension of the cell body that lacks Nissl substance.

        3. The initial segment is adjacent to the axon hillock.

            a. The membrane of the initial segment contains numerous voltage-sensitive sodium ion channels.

            b. The initial segment is the "trigger zone" of an axon where conduction of electrical activity as an action potential is initiated.

            c. If the axon is myelinated, the myelin sheath begins at the initial segment.

        4. The cytoplasm of the entire axon lacks free polysomes, Nissl substance, and Golgi apparatus but contains mitochondria and smooth endoplasmic reticulum.

    H. Three common **histological techniques** are used in staining neurons and their processes (Figure 2–4).

        1. Nissl stains are basophilic dyes, which stain the membrane-bound components of a neuron, in particular those that contain Nissl substance.

            a. Nissl stains demonstrate the nucleus, cell body, and proximal parts of the dendrites of neurons.

            b. Nissl stains can be used to illustrate the size and number of neurons in a nucleus in the CNS.

            c. Axons and the distal parts of dendrites will not be stained by Nissl stains.

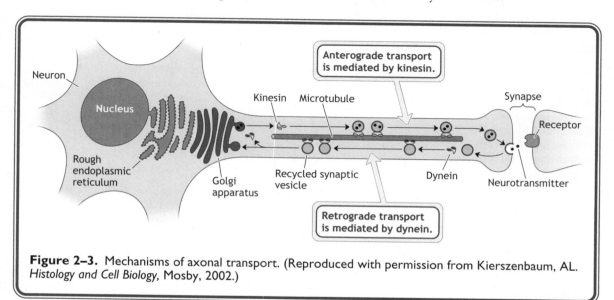

Figure 2–3. Mechanisms of axonal transport. (Reproduced with permission from Kierszenbaum, AL. *Histology and Cell Biology,* Mosby, 2002.)

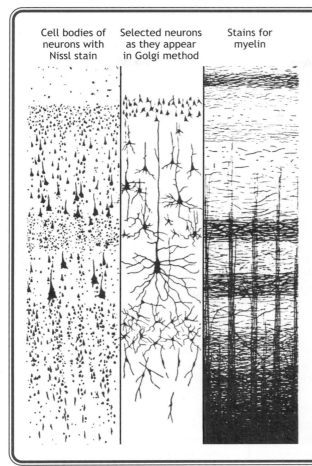

Cell bodies of neurons with Nissl stain

Selected neurons as they appear in Golgi method

Stains for myelin

Figure 2–4. Histological techniques that demonstrate elements of neural tissue.

    2. Fiber or myelin (e.g., Weigert's) stains stain myelinated axons in nerves or tracts. Nuclei containing the cell bodies of neurons are not stained.

    3. Reduced silver methods, such as the Golgi method, react with all parts of a limited number of neurons and are useful for studying the branching pattern of dendrites.

II. Axonal Transport

    A. Anterograde axonal transport moves proteins and membranes that are synthesized in the cell body through the axon to the synaptic terminal (see Figure 2–3).

      1. In **fast anterograde transport,** there is a rapid (100–400 mm/day) movement of materials from the cell body to the axon terminal.

        a. Fast anterograde transport is dependent on **kinesin,** an ATPase, which acts as the motor molecule.

        b. Kinesin links the membrane of the element being transported to microtubules, which act like railroad tracks in the transport process.

        c. Fast anterograde transport delivers precursors of peptide neurotransmitters to synaptic terminals.

2. In **slow anterograde transport,** there is a slow (1–2 mm/day) anterograde movement of soluble cytoplasmic components.
   a. Cytoskeletal proteins, enzymes, and precursors of small molecule neurotransmitters are transported to synaptic terminals by slow anterograde transport.
   b. Slow transport is not dependent on microtubules or ATPase motor molecules.
3. Anterograde axonal **transport of radiolabeled amino acids,** such as tritiated leucine, is used in investigations to determine the sites of terminations of axons of specific populations of neurons.
   a. Radiolabeled amino acids are injected in the vicinity of neuron cell bodies, the neurons take up the amino acid and incorporate it into proteins, and the proteins are transported to axon terminals.
   b. Autoradiography demonstrates the presence of the radiolabeled proteins in the axon terminals.

NEUROPATHIES AND AXONAL TRANSPORT

- *Disruption of fast anterograde transport* may result in an *axonal polyneuropathy* and may be caused by anoxia, which affects mitochondrial oxidative phosphorylation, or by anticancer agents, such as colchicine and vinblastine, which depolymerize microtubules.
- In patients with *diabetes,* hyperglycemia results in an alteration of proteins that form microtubules, which may disrupt axonal transport. Patients may develop axonal polyneuropathies in long axons in nerves that produce a *"glove-and-stocking" pattern* of altered sensation and pain in the feet and then in the hands.

B. **Retrograde axonal transport** returns intracellular material from the synaptic terminal to the cell body to be recycled or digested by lysosomes.
  1. Retrograde transport also **uses microtubules and is slower than anterograde transport** (60–100 mm/day).
  2. Retrograde transport is **dependent on dynein, an ATPase,** which acts as the retrograde motor molecule.
  3. Retrograde transport also **permits communication between the synaptic terminal and the cell body** by transporting trophic factors emanating from the postsynaptic target or in the extracellular space.
  4. **Retrograde axonal transport of horseradish peroxidase** (HRP) is used in investigations to determine the locations of neuron cell bodies.
     a. HRP is injected in the vicinity of axon terminals, and the axons take up the HRP and transport it back to the neuron cell bodies to be degraded by lysosomes.
     b. The labeled neuron cell bodies are reacted histochemically with a chromogen to determine which cells contain the HRP reaction product (Figure 2–5).

AXONAL TRANSPORT AND NEUROLOGICAL DISORDERS

- The *polio, herpes, and rabies viruses and tetanus toxins* are taken up and retrogradely transported by axons that innervate skeletal muscle.
- Patients with *polio and rabies may have weakness of skeletal muscles* innervated by the affected neurons. Those with *tetanus have continuous contractions and spasms of axial and limb muscles and spasms of muscles innervated by certain cranial nerves.*
- *Herpes is taken up and retrogradely transported* in sensory fibers and remains dormant in sensory ganglia.

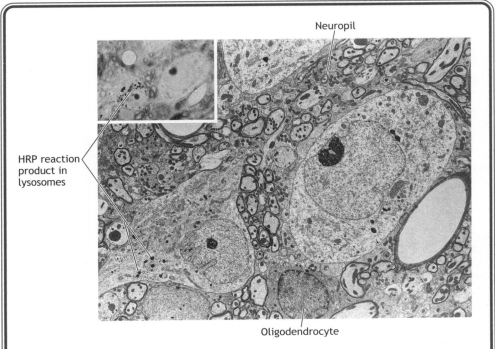

Figure 2–5. Neuron, on left in light micrograph (inset) and in electron micro-graph, contains lysosomes reacted for presence of horseradish peroxidase following retrograde axonal transport. Neuron on right is unlabeled (White, JS and Warr, WB, unpublished).

• *The rabies and herpes viruses are also transported in the anterograde direction. When activated, the herpes virus results in painful vesicles corresponding to the distribution of the affected sensory fibers. Rabies is transported to salivary glands and is transmitted by saliva after an animal bite. A skunk bite is the most common cause of rabies transmission in the United States.*

III. Synaptic Terminals

  A. **Synaptic terminals** are situated at the distal ends of axons, where neural activity is unidirectionally transmitted to a postsynaptic target at a synapse (Figure 2–6A and B).

  B. **Chemical neurotransmission** is mediated mainly by the release of neurotrans-mitter from the synaptic terminal.

  C. The synaptic terminal contains clusters of synaptic vesicles that are attached to docking proteins in the presynaptic density.

  D. At the terminal, the arrival of the action potential initiates a **series of calcium ion-dependent events.**

   1. Ca^{2+} channels in the synaptic terminal membrane open, and calcium enters and accumulates near the synapse.

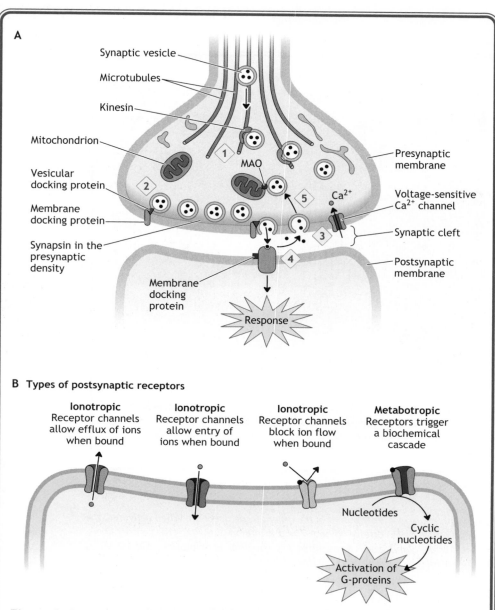

A

Synaptic vesicle

Microtubules

Kinesin

Mitochondrion

Vesicular docking protein

Membrane docking protein

Synapsin in the presynaptic density

Membrane docking protein

Response

MAO

Presynaptic membrane

Voltage-sensitive Ca²⁺ channel

Synaptic cleft

Postsynaptic membrane

B Types of postsynaptic receptors

Ionotropic
Receptor channels allow efflux of ions when bound

Ionotropic
Receptor channels allow entry of ions when bound

Ionotropic
Receptor channels block ion flow when bound

Metabotropic
Receptors trigger a biochemical cascade

Nucleotides

Cyclic nucleotides

Activation of G-proteins

Figure 2–6. A and B: Types of postsynaptic receptors and features of a synapse. (**A**: Reproduced with permission from Kierszenbaum, AL. *Histology and Cell Biology*, Mosby, 2002.)

1. Neurotransmitters are transported to the synaptic terminal by kinesin-mediated anterograde axonal transport
2. Docking proteins in vesicle membrane attach to membrane docking proteins in presynaptic density
3. Arrival of action potential causes Ca²⁺ to enter terminal and induces exocytosis of neurotransmitter into synaptic cleft
4. Neurotransmitter binds with a postsynaptic receptor
5. Neurotransmitter is degraded in the cleft or is taken up and degraded by monoamine oxidase (MAO).

2. Ca^{2+} influx induces synaptic vesicles to release neurotransmitter into the synaptic cleft.

3. The neurotransmitter binds with a variety of receptors in the postsynaptic membrane.

E. The **effect of the neurotransmitter** depends on the action of the receptors with which the neurotransmitter binds or interacts (see later discussion).

F. At **gap junctions,** neural activity is transmitted from one neuron to another by direct electrical coupling through **connexon channels** that link adjacent cells.

1. Gap junctions permit rapid bidirectional transmission between interconnected cells without using chemical neurotransmitters.

2. Gap junctions permit the intercellular passage of inorganic cations, like calcium, and anions, ATP, and second messengers, such as cyclic adenosine monophosphate (cAMP), and peptides of a molecular weight less than 1000 daltons.

GAP JUNCTIONS

Gap junctions close in damaged neurons and decouple the damaged cells from other neurons in response to a decrease in intracellular pH or elevated intracellular Ca^{2+} (as in glutamate-induced excitotoxicity; see later discussion).

IV. Chemical Neurotransmission

A. **Chemical neurotransmission** is a multistep process that includes synthesis, storage, release, receptor binding, and reuptake or inactivation of a neurotransmitter or its precursors; a synaptic terminal may contain and use 1 or more chemical neurotransmitters (see Figures 2–3, 2–6A and B).

B. **Small molecule neurotransmitters** are fast-acting biogenic amines or amino acid neurotransmitters that are synthesized in axon terminals.

1. The enzymes for synthesizing small molecule neurotransmitters and precursors of the neurotransmitters are produced by free polysomes in the neuron cell body.

2. Enzymes and their precursors are transported to the axon terminal by slow anterograde axonal transport.

3. In the axon terminal, the enzyme converts the precursor into the neurotransmitter, which is then incorporated into synaptic vesicles.

4. Acetylcholine (ACh) is a small molecule excitatory neurotransmitter used by axon terminals in both the PNS and the CNS.

   a. In the PNS, ACh is used by all preganglionic autonomic neurons, all postganglionic parasympathetic neurons, and postganglionic sympathetic neurons that innervate eccrine sweat glands.

   b. In the PNS, ACh is used by axons of motor neurons at neuromuscular junctions in skeletal muscle.

   c. In the CNS, ACh is used by neurons in the basal ganglia, in the reticular formation near the pons-midbrain junction, in the ventral forebrain (basal nucleus of Meynert), and in the septal nuclei.

   d. **Cholinergic neurons** in the brainstem near the pons-midbrain junction **form part of the ascending arousal system** that projects to the diencephalon and cortex and helps maintain an awake state.

e. Cholinergic neurons **initiate rapid eye movement (REM) sleep** and facilitate the activity of hippocampal neurons in memory consolidation.

f. ACh is synthesized in synaptic terminals from acetyl coenzyme A (CoA) and choline, and, after release, is degraded in the synaptic cleft by acetylcholinesterase. Choline is taken up and recycled from the synaptic cleft.

DISORDERS ASSOCIATED WITH CHOLINERGIC NEURONS

- *Organophosphates,* such as mustard gas and insecticides, bind irreversibly to acetylcholinesterase at the neuromuscular junction. ACh accumulates in the synaptic cleft and causes prolonged depolarization of skeletal muscle, making it less sensitive to additional ACh release. Death may result from respiratory paralysis.
- In the CNS, cholinergic neurons in the ventral forebrain (basal nucleus of Meynert) are a main type of neuron that degenerates in patients with Alzheimer's disease.

5. Glutamate is a neurotransmitter used by virtually all excitatory neurons in the CNS and is the neurotransmitter of 50% of the neurons in the CNS.

a. Glutamate is synthesized from the amino acid glutamine in the axon terminal and, after release, is rapidly taken up by transporters in presynaptic terminals and in astrocytes.

b. Astrocytes convert glutamate to glutamine; glutamine is released by the astrocytes, taken up, and reused by the presynaptic terminal.

GLUTAMATE-INDUCED EXCITOTOXICITY

- Ischemic neurons in stroke patients may have *glutamate-induced excitotoxicity,* in which glutamate accumulates in the synaptic cleft, because the uptake process is energy dependent and is slowed by oxygen deprivation. Epilepsy, characterized by multiple seizures that arise from excessive neuronal activity in the cerebral cortex, may also cause glutamate-induced excitotoxicity of gamma-aminobutyric acid (GABA) neurons, increasing the likelihood of additional seizures.
- Excessive levels of glutamate result from an increase in the influx of Ca^{2+} into the postsynaptic cell at N-methyl-D-aspartate (NMDA) receptors. The influx of Ca^{2+} may result in the production of free radicals in the postsynaptic cell, which damages the membranes of its organelles and the plasma membrane. Disruption of the plasma membrane results in an influx of water, which leads to swelling and eventual lysis of the cell.
- Blockage of the NMDA Ca^{2+} receptors that respond to glutamate may be effective in reducing neuronal loss in stroke patients.

6. GABA and glycine are major inhibitory CNS transmitters; GABA is used by about 30% of all CNS neurons, and glycine is used mainly by neurons in the brainstem and spinal cord.

a. Both GABA and glycine prevent prolonged excitation in many neuronal systems.

b. Serial populations of GABA neurons are used by the basal ganglia to disinhibit nuclei in the diencephalon.

c. GABA is synthesized from glutamate in the synaptic terminal and, after release, is transported back into the terminal or degraded into succinate.

d. Glycine is synthesized from serine and, after release, is transported back into the presynaptic terminal.

ANTIANXIETY MEDICATIONS

*Benzodiazepines, such as diazepam (Valium) and chlordiazepoxide (Librium), are effective **antianxiety drugs that act by enhancing the effects of GABA neurons.** Benzodiazepines and barbiturates also act as antiepileptic drugs by enhancing GABA inhibition.*

7. **Dopamine** is a catecholamine synthesized from the amino acid tyrosine; tyrosine hydroxylase, the rate-limiting enzyme in catecholamine synthesis, catalyzes the conversion to dopamine.
   a. Dopaminergic neurons are found in the substantia nigra, pars compacta, the ventral tegmental area of the midbrain, and the arcuate nucleus of the hypothalamus.
   b. Nigral and tegmental axons form mesostriatal, mesocortical, and meso-limbic dopamine pathways that project to the striatum of the basal ganglia and to frontal and limbic lobes of cerebral cortex.
   c. The arcuate nuclei of the hypothalamus use dopamine to regulate the release of prolactin from the anterior pituitary.

DOPAMINERGIC NEURONS, PARKINSON'S DISEASE, AND SCHIZOPHRENIA

- *Selective loss of the dopaminergic neurons in the substantia nigra, pars compacta, results in Parkinson's disease, a motor disorder of the basal ganglia.*
- *Changes in the levels of dopamine in the mesolimbic and mesocortical circuits may result in the positive and negative symptoms of schizophrenia, respectively.*
- *Cocaine and amphetamines are addictive substances that act by increasing dopamine levels, particularly in the mesolimbic pathway.*

8. **Norepinephrine** is a catecholamine that is converted from dopamine and is the only small neurotransmitter synthesized inside synaptic vesicles at the synaptic terminal instead of in the cytoplasm of the terminal.
   a. Noradrenergic neurons are found mainly in the locus ceruleus in the midbrain and in the lateral tegmental nucleus of the medulla.
   b. Noradrenergic neurons project to all parts of the CNS; norepinephrine is also used by most postganglionic sympathetic neurons.
   c. Changes in the levels of norepinephrine in the CNS influence sleep, wakefulness, and states of arousal.
9. **Epinephrine** is converted in small amounts from norepinephrine in the CNS.
   a. Adrenergic neurons in the CNS are found mainly in the medulla.
   b. Most epinephrine is synthesized and released outside the CNS in the adrenal medulla.
   c. After release, all catecholamines are taken up by axon terminals and degraded by either monoamine oxidase (MAO) or catechol-o-methyltransferase (COMT).
10. **Serotonin** (5-hydroxytryptamine [5-HT]) is an indolamine synthesized from tryptophan; tryptophan hydroxylase is the rate-limiting synthetic enzyme.
    a. Serotonin is an excitatory neurotransmitter that influences neuronal circuits, which function in sleep, arousal, eating, and circadian rhythms.
    b. Serotonergic neurons are found in the raphe nuclei near the midline of the brainstem; their axons project to the cerebral cortex and basal ganglia, diencephalon, and spinal cord.

DEPRESSION

- *Changes in the synthesis, release, or activation of catecholamines and serotonin may cause depression,* the most common unipolar mood disorder, or *bipolar disorder,* during which patients experience both depression and euphoric (manic) episodes.
- *Antidepressant drugs include MAO inhibitors,* which limit the reuptake of catecholamines. Drugs such as fluoxetine (Prozac) (for moderate depression) selectively block the uptake of serotonin without affecting the reuptake of other catecholamines, and tricyclic compounds (for severe depression) inhibit the uptake of both serotonin and norepinephrine.

11. **Histamine** is an excitatory neurotransmitter used by neurons in the posterior hypothalamus. It has effects similar to acetylcholine and norepinephrine in facilitating arousal and an awake state.
    a. Histamine is synthesized from histidine and, after its release, is transported back into axon terminals and metabolized by MAO.
    b. Histamine is also released from mast cells in allergic reactions and in response to tissue damage.
12. **ATP** is an excitatory cotransmitter found in all synaptic vesicles and is released with other small molecule neurotransmitters.
    a. ATP is synthesized from adenosine diphosphate by oxidative phosphorylation.
    b. ATP is used by motor neurons in the spinal cord.
    c. ATP is found in autonomic and sensory ganglia.
    d. ATP is a neurotransmitter that may be intercellularly transmitted by way of gap junctions.
C. **Large molecule neurotransmitters are slow-acting neuroactive peptides** synthesized in the rough endoplasmic reticulum.
    1. **Neuropeptides** or their precursor proteins are transported from the rough endoplasmic reticulum to the Golgi apparatus, where they are packaged into vesicles and then transported to the synaptic terminal by fast anterograde axonal transport.
    2. **Many neuropeptides act as cotransmitters,** are released with small molecule neurotransmitters, and are degraded by proteases.
    3. There are more than 50 **neuroactive peptides** grouped in 5 general categories: brain-gut peptides, opioid peptides, pituitary peptides, hypothalamic-releasing hormones, and others that do not fit into a specific category.
    4. **Substance P** is the best known brain-gut peptide.
       a. Substance P is found in neurons in the cerebral cortex, basal ganglia, hippocampus, and gastrointestinal tract.
       b. Substance P is used by axons of neurons that respond to pain and project into the spinal cord in dorsal roots of spinal nerves.
    5. There are 3 classes of about 20 **opioid peptides,** including enkephalins, endorphins, and dynorphins, that act mainly as analgesics; their effects are mimicked by narcotic drugs.
       a. Enkephalins are used by neurons in the dorsal horn of the spinal cord in the suppression of pain.
       b. Opioid peptides are also used by neurons in the periaqueductal gray matter of the midbrain in the suppression of pain.

6. Pituitary neuropeptides include the hormones oxytocin and vasopressin, which are produced by the neurons in the hypothalamus and released from axon terminals of those neurons into the posterior pituitary (neurohypophysis).

    7. Hypothalamic-releasing and hypothalamic-inhibiting hormones are synthesized by neurons in the arcuate and periventricular nuclei of the hypothalamus and influence the release of anterior pituitary peptide hormones; dopamine acts as an inhibitory factor that regulates the release of prolactin.

V. Neurotransmitter Receptors

  A. At chemical synapses, released neurotransmitter binds with specific receptors that alter the permeability of the postsynaptic membrane (see Figures 2–6A and B).

    1. The postsynaptic response may be excitatory, resulting in an **excitatory postsynaptic potential** (EPSP), which increases the likelihood of the generation of an action potential in a postsynaptic axon.

      a. EPSPs result from a net inward depolarizing current caused by increased Na^+ and Ca^{2+} conductance at postsynaptic receptors.

      b. Glutamate and acetylcholine are the common CNS excitatory neurotransmitters that generate EPSPs.

    2. The response may be inhibitory, resulting in an **inhibitory postsynaptic potential** (IPSP), which decreases the likelihood of the generation of an action potential.

      a. IPSPs result from a net outward hyperpolarizing current caused by increased outward conductance of K^+ and inward conductance of Cl^- at postsynaptic receptors.

      b. GABA and glycine are the common CNS inhibitory neurotransmitters that generate IPSPs.

  B. Ionotropic receptors typically consist of 5 protein subunits that form or are linked to an ion channel (see Figure 2–6B).

    1. Binding of a neurotransmitter to an ionotropic receptor results in a direct increase or decrease in the ion conductance of the channel.

    2. Ionotropic receptors are located on postsynaptic elements where a fast response to the presynaptic signal is required.

    3. Nicotinic acetylcholine receptors are ionotropic receptors in the membranes at the neuromuscular junctions in skeletal muscle, on postganglionic autonomic neurons, and in the CNS.

      a. When ACh binds to the receptor, Na^+ ions flow into the cell, and K^+ ions flow out of the cell through the channel.

      b. These receptors are so named because nicotine binds to the receptor and mimics the effects of ACh.

DISORDERS OF NEUROMUSCULAR TRANSMISSION: BOTULINUM TOXIN, MYASTHENIA GRAVIS, CURARE, α-BUNGAROTOXIN (SNAKE VENOM), AND LAMBERT-EATON SYNDROME (FIGURE 2–7)

• ***Botulinum toxin*** *prevents the release of ACh by binding to the presynaptic membrane at neuromuscular junctions and at preganglionic autonomic terminals. Both skeletal and smooth muscles are affected. Patients have weakness initially in muscles innervated by cranial nerves and then weakness in*

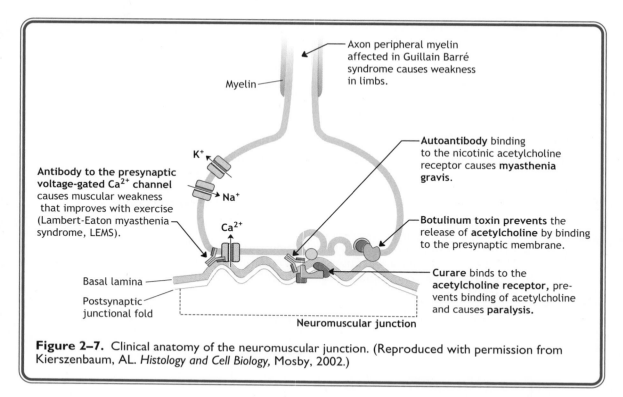

Myelin

Axon peripheral myelin affected in Guillain Barré syndrome causes weakness in limbs.

K^+

Na^+

Antibody to the presynaptic voltage-gated Ca^{2+} channel causes muscular weakness that improves with exercise (Lambert-Eaton myasthenia syndrome, LEMS).

Ca^{2+}

Autoantibody binding to the nicotinic acetylcholine receptor causes **myasthenia gravis.**

Botulinum toxin prevents the release of **acetylcholine** by binding to the presynaptic membrane.

Curare binds to the **acetylcholine receptor,** prevents binding of acetylcholine and causes **paralysis.**

Basal lamina

Postsynaptic junctional fold

Neuromuscular junction

Figure 2–7. Clinical anatomy of the neuromuscular junction. (Reproduced with permission from Kierszenbaum, AL. *Histology and Cell Biology,* Mosby, 2002.)

---

the limbs. Patients also have a dry mouth and abdominal cramping, vomiting, and diarrhea because of an absence of peristalsis.

- In patients with **myasthenia gravis** (MG), antibodies bind to the nicotinic cholinergic receptors, and the receptor complex is engulfed and destroyed, reducing the number of available receptors. Patients have weakness mainly in skeletal muscles innervated by cranial nerves. Ninety percent of diagnosed patients have ocular weakness and a bilateral ptosis. Most patients with MG have a thymoma or thymic hyperplasia.

- **Curare and α-bungarotoxin** also bind to the nicotinic cholinergic receptors and block neuromuscular transmission, producing effects similar to those seen in patients with MG.

- In patients with **Lambert-Eaton myasthenic syndrome** (LEMS), an underlying neoplasm is usually present. Antibodies raised against the neoplasm bind with presynaptic calcium ion channels and disrupt neuromuscular transmission. Skeletal muscles in the limbs are mainly affected. Sustained contractions of affected muscles lead to an increase in strength.

4. There are 3 classes of ionotropic glutamate receptors named for the compounds that activate them.
  a. 2–(Aminomethyl) phenylacetic acid (AMPA) and kainate receptors consist of 6 different proteins that form the 5 subunits around an ion channel.
  b. Activation of these receptors results in a rapid increase in the conductance of Na^+ and K^+ ions.
  c. The NMDA receptor consists of 5 subunits.
  d. Activation of the NMDA receptor results in an increase in the conductance of Na^+ ions and Ca^{2+} ions and the production of EPSPs.

5. **Long-term potentiation** (LTP) occurs at NMDA glutamate receptors (best studied in the hippocampus) and may be a mechanism used by CNS neurons in learning and memory.
   a. When the postsynaptic cell membrane is at its resting potential, glutamate does not bind with NMDA receptors, and Ca^{2+} ions are prevented from entering the cell because Mg^{2+} ions block the channel.
   b. LTP occurs when 2 simultaneous presynaptic excitatory inputs depolarize the postsynaptic membrane and remove the Mg^{2+} block of the NMDA receptor.
   c. An influx of Ca^{2+} occurs, which results in increased protein synthesis, an increase in the strength of the NMDA receptor, and production of more NMDA receptors.
   d. The increased postsynaptic response may also result in the release of nitric oxide from the postsynaptic terminal; nitric oxide acts as a retrograde signal that directs the presynaptic terminal to release more glutamate in response to additional action potentials.
6. **Long-term depression** (LTD) (best studied in the cerebellum) results from synaptic transmission between excitatory parallel fibers (axons of granule cells) and Purkinje cells.
   a. Activated parallel fibers release glutamate, which binds with AMPA receptors and causes a brief depolarization of Purkinje cells.
   b. AMPA receptor binding generates second messengers that cause Ca^{2+} to be released from the endoplasmic reticulum of the Purkinje cell and pumped out of the cell.
   c. The Ca^{2+} release causes LTD by making the parallel fiber synapses less effective.
7. There are **2 classes of GABA receptors:** $GABA_A$ receptors are ionotropic; $GABA_B$ receptors are metabotropic (see later discussion).
   a. $GABA_A$ receptors consist of 5 different protein subunits.
   b. Activation of the $GABA_A$ receptor results in a rapid increase of Cl^- influx into the cell, making the membrane less responsive to EPSPs from other synapses.
   c. $GABA_A$ receptor effects are enhanced by alcohol, steroid anesthetics, and diazepam (Valium).
8. **Glycine receptors** are similar in structure and function to $GABA_A$ receptors.

STRYCHNINE ANTAGONIST

Strychnine is an antagonist *that binds to and blocks glycine receptors; patients have spasms similar to those infected with tetanus toxin, but cranial nerve innervated muscles are spared.*

C. **Metabotropic receptors** are separate from ion channels and cause movement of ions through ion channels only after several metabolic steps (see Figure 2–6B).
   1. Most metabotropic receptors are known as G protein-coupled receptors because receptor binding activates intermediate G proteins.
      a. G proteins interact with adjacent ion channels or activate enzymes that produce metabolites that act as second messengers.
      b. Cyclic nucleotides, calcium, and nitric acid are examples of second messengers that may directly alter ion conductance at ion channels and activate cascades that change the biochemical state of the cell.

c. Neuropeptides and hormones that act at metabotropic receptors activate receptor tyrosine kinases instead of using G protein-coupled receptors and second-messenger systems.

    2. Metabotropic receptors consist of a single polypeptide with 7 subunits; the nervous system contains more than 250 different metabotropic receptors.

    3. Metabotropic receptors mediate a slower postsynaptic response that has a longer duration than responses generated by ionotropic receptors.

    4. There are 5 types of muscarinic metabotropic acetylcholine receptors (mAChRs): m-1 through m-5.

        a. The main mAChRs in the brain are m-1, m-3, and m-4; m-2 mAChRs are found in target tissues innervated by autonomic axons.

        b. Muscarinic receptors are found in both presynaptic and postsynaptic membranes and mediate excitatory or inhibitory effects.

    5. There are 3 families of adrenergic receptors in the nervous system: α_1, α_2, and β receptors; each family also has 3 additional subclasses.

        a. Adrenergic receptors are located in the membranes of presynaptic terminals.

        b. Adrenergic receptors act mainly as autoreceptors that bind released neurotransmitter and inhibit subsequent release of that neurotransmitter.

    6. The 2 main classes of metabotropic dopamine receptors, D_1-like and D_2-like, are found in neurons in the basal ganglia and cortex.

        a. D_1-receptor binding produces excitatory postsynaptic effects.

        b. D_2-receptor binding produces inhibitory postsynaptic effects.

    7. There are 4 classes of serotonin receptors: 5-HT1 through 5-HT4; all but 5-HT3 are metabotropic receptors.

    8. $GABA_B$ receptors are metabotropic receptors that mediate inhibition of neurotransmitter release by increasing K^+ conductance and decreasing Ca^{2+} conductance at the presynaptic terminal.

    9. There are 3 classes of metabotropic glutamate receptors (mGluRs) with widespread distribution throughout the CNS.

VI. Glial and Supporting Cells in the CNS and PNS

  A. The **supporting, or glial, cells of the CNS** are small cells, which differ from neurons.

    1. Supporting cells have only 1 kind of process and do not form chemical synapses.

    2. Unlike neurons, supporting cells readily divide and proliferate; gliomas are the most common type of primary tumor of the CNS.

    3. Supporting cells outnumber neurons by a factor of 10.

  B. **Astrocytes are the most numerous glial cells** in the CNS and have large numbers of radiating processes (Figures 2–8 and 2–9).

    1. Astrocytes provide the structural support, or scaffolding, for the CNS and contain large bundles of intermediate filaments that consist of **glial fibrillary acidic protein** (GFAP).

    2. Astrocytes have uptake systems that remove the neurotransmitter glutamate and K^+ ions from the extracellular space.

    3. Astrocytes have foot processes that contribute to the blood-brain barrier by forming a glial-limiting membrane.

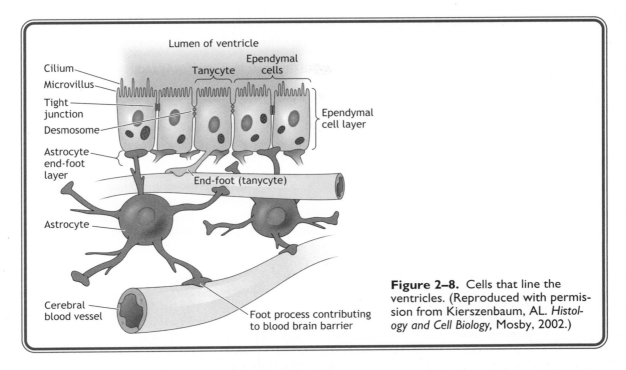

Figure 2–8. Cells that line the ventricles. (Reproduced with permission from Kierszenbaum, AL. *Histology and Cell Biology,* Mosby, 2002.)

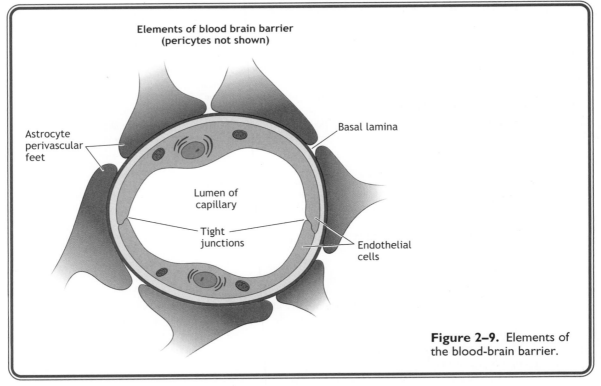

Figure 2–9. Elements of the blood-brain barrier.

4. Astrocytes induce the formation of tight junctions between endothelial cells of cerebral blood vessels at the blood-brain barrier (see later discussion).

5. Astrocytes hypertrophy and proliferate after an injury to the CNS; they fill up the extracellular space left by degenerating neurons by forming an astroglial scar.

6. Astrocytes phagocytose neuronal debris after injury.

7. Astrocytes are linked by gap junctions.

8. **Radial glia** are precursors of astrocytes that guide neuroblast migration during CNS development.

    a. Radial glia have processes that extend from the ventricular to the pial surface of the brain.

    b. Radial glia provide a scaffolding that enables immature neurons to migrate away from the ventricular zone to the intermediate or marginal zones of the neural tube.

9. **Müller cells** are astrocytic glial cells in the retina, and **pituicytes** are specialized astrocytes in the neurohypophysis.

C. **Microglia cells** are the smallest glial cells in the CNS.

1. Unlike the rest of the CNS neurons and glia, which are derived from neuroectoderm, microglia are derived from bone marrow monocytes and enter the CNS after birth.

2. Microglia provide a link between cells of the CNS and the immune system.

3. Microglia proliferate and migrate to the site of a CNS injury and phagocytose neuronal debris after injury.

4. Pericytes are microglia that contribute to the blood-brain barrier.

MICROGLIA AND CNS DISORDERS

Microglia determine the chances of survival of a CNS tissue graft and are the cells in the CNS that are targeted by the HIV-1 virus in patients with AIDS. The affected microglia may produce cytokines that are toxic to neurons.

CNS microglia that become phagocytic in response to neuronal tissue damage may secrete toxic free radicals. Accumulation of free radicals, such as superoxide, may lead to disruption of the calcium homeostasis of neurons. The degenerative changes seen in patients with Alzheimer's disease and in those with MS may be caused by free radical damage resulting from a progressive inflammatory response caused by microglia.

D. **Oligodendrocytes** have few processes.

1. Oligodendrocytes form myelin for axons in the CNS (see Figure 2–7). Each of the processes of the oligodendrocyte can myelinate individual segments of many axons.

2. Unmyelinated axons in the CNS are not ensheathed by oligodendrocyte cytoplasm.

E. **Schwann cells** are the supporting cells of the PNS.

1. Schwann cells are derived from neural crest cells.

2. Schwann cells form the myelin for axons and processes in the PNS. Each Schwann cell forms myelin for only a single internodal segment of a single axon (Figures 2–7 and 2–10A and B).

3. Unmyelinated axons in the PNS are enveloped by the cytoplasmic processes of a Schwann cell.

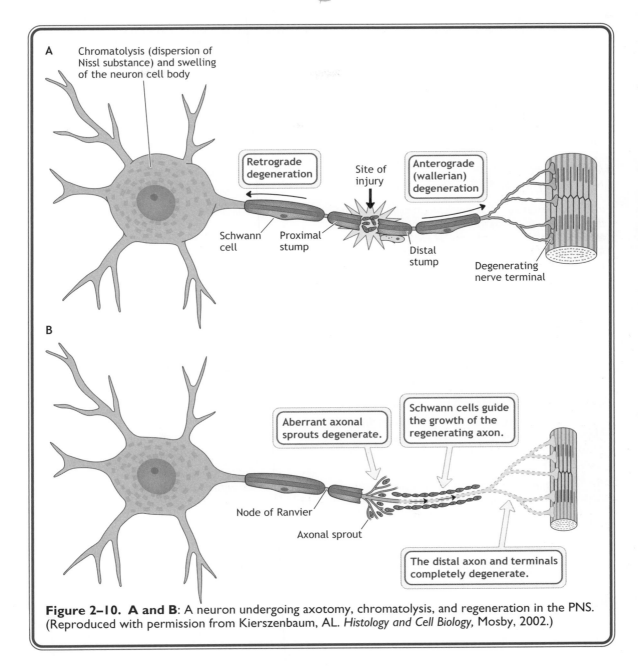

Figure 2–10. A and B: A neuron undergoing axotomy, chromatolysis, and regeneration in the PNS. (Reproduced with permission from Kierszenbaum, AL. *Histology and Cell Biology,* Mosby, 2002.)

4. Schwann cells act as phagocytes and remove neuronal debris in the PNS after injury.
5. A **node of Ranvier** is the region between adjacent myelinated segments of axons in the CNS and the PNS (see Figure 2–10B).
   a. The membranes at nodes of Ranvier have numerous sodium channels.

b. In all myelinated axons, nodes of Ranvier are sites that permit action potentials to jump from node to node (saltatory conduction).

    c. Saltatory conduction dramatically increases the conduction velocity of impulses in myelinated axons.

NODES OF RANVIER AND DIABETES

In most patients with diabetes, sorbitol, which is normally metabolized from glucose, accumulates in nerves and impairs conductance of ion channels and saltatory conduction at nodes of Ranvier.

IMMUNE-RELATED DEMYELINATING DISEASES, MULTIPLE SCLEROSIS, AND GUILLAIN-BARRÉ SYNDROME

- *In **multiple sclerosis** (MS), myelin formed by oligodendrocytes undergoes an inflammatory reaction that impairs impulse transmission in axons in the CNS.*
- ***MS is marked by the presence of plaques,** which are sharply demarcated areas of demyelination. MS plaques tend to form in axons that course near the surfaces of the lateral ventricles, in the floor of the fourth ventricle, or near the pial surfaces of the brainstem or spinal cord.*
- *In patients with MS, **multiple lesions appear over time,** but the signs and symptoms may undergo exacerbation and remission; commonly, 2 or more CNS sensory or motor neural systems are affected in separate attacks.*
- ***The only cranial nerve or spinal nerve affected by MS is the optic nerve** (cranial nerve [CN] II) because all the myelin sheaths of its axons are formed by oligodendrocytes. MS is the most common cause of optic neuritis.*
- *In patients with MS, cerebrospinal fluid (CSF) contains elevated γ-globulin and T lymphocytes but has normal levels of glucose.*
- *In Guillain-Barré syndrome, myelin formed by Schwann cells in the PNS undergoes an acute inflammatory reaction after a respiratory or gastrointestinal illness. This reaction impairs or blocks impulse transmission of axons in the PNS resulting in a polyneuropathy.*
- *Motor axons are always affected, producing weakness in the limbs. Weakness of cranial nerve innervated muscles (most commonly those innervated by CNs VI and VII) or respiratory muscles may be seen. Sensory deficits are mild or absent.*

    F. Ependymal cells line the ventricles in the adult brain (see Figure 2–8).

        1. Some ependymal cells differentiate into choroid epithelial cells, forming part of the choroid plexus, which produces CSF.

        2. Ependymal cells are ciliated; ciliary action helps circulate CSF.

        3. Tanycytes are specialized ependymal cells that have basal cytoplasmic processes in contact with blood vessels; these processes may transport substances between a blood vessel and a ventricle.

GLIAL NEOPLASMS

- *Astrocytomas are the most common form of primary neoplasm in the CNS. In adults, 70% of neoplasms are supratentorial. The most common neoplasm is a **glioblastoma multiforme,** a malignant astrocytoma. In children, 70% of neoplasms are infratentorial, and the most common is a **benign cerebellar astrocytoma.***
- ***Oligodendrogliomas, ependymomas, and microgliomas** occur less frequently. Cells of the pia and arachnoid and Schwann cells may give rise to meningiomas and schwannomas, respectively.*
- *Primary tumors of the CNS (50% of CNS tumors) rarely metastasize outside the CNS. Secondary metastases (50% of CNS tumors) may invade the CNS from primary sites outside of the nervous system.*
- *Glial neoplasms are space-occupying lesions that may result in an increase in intracranial pressure.*

VII. The Blood–Brain Barrier

A. The **blood-brain barrier** restricts access of microorganisms, proteins, cells, and drugs to the nervous system.

B. The **blood-brain barrier consists of capillary endothelial cells, an underlying basal lamina, astrocytes, and pericytes** (see Figure 2–9).

    1. Cerebral capillary endothelial cells and their intercellular junctions are the most important elements of the blood-brain barrier.

        a. The cell membranes of capillary endothelial cells are joined by high-resistance intercellular tight junctions.

        b. Capillary endothelial cells lack fenestrations, which are typical of endothelial cells elsewhere in the body.

    2. Astrocytes and pericytes are found at the blood-brain barrier outside the basal lamina.

        a. Astrocytes induce the capillary endothelial cells to form the blood-brain barrier and maintain the tight junctions between them.

        b. Astrocytes have processes with "end feet" that cover more than 95% of the basal lamina adjacent to the capillary endothelial cells (see Figure 2–8).

        c. Pericytes are microglial cells that are interposed between the astrocyte foot processes and the basal lamina.

        d. Pericytes control proliferation and repair of endothelial cells, regulate vessel tone, and secrete elements of the basal lamina.

        e. Pericytes show phagocytic activity after injury to the blood-brain barrier.

C. **Substances cross the blood-brain barrier into the CNS by diffusion, by selective transport, and via ion channels.**

    1. Oxygen and carbon dioxide are lipid-soluble gases that readily diffuse across the blood-brain barrier.

    2. Glucose, amino acids, and vitamins K and D are water-soluble substances that are selectively transported across the blood-brain barrier.

    3. Sodium and potassium ions move across the blood-brain barrier through ion channels.

DRUGS OF ADDICTION AND THE BLOOD-BRAIN BARRIER

Heroin, ethanol, and nicotine are lipid-soluble compounds that readily diffuse across the blood-brain barrier.

NEOPLASMS AND THE BLOOD-BRAIN BARRIER

Capillaries that supply primary neoplasms of the CNS are fenestrated and form an incomplete blood-brain barrier. These "leaky" capillaries cause vasogenic edema, an increase in CNS interstitial fluid levels. Tumors of the CNS readily absorb contrast materials through the leaky capillaries, rendering them detectable by imaging techniques.

VIII. Response of Axons to Destructive Lesions (Severing an Axon or Axotomy) or Irritative Lesions (Compression of an Axon)

A. An axon that is severed in either the CNS or PNS undergoes **anterograde or wallerian degeneration** distal to the cut (see Figure 2–10A and B).

1. The closer the destructive lesion is to the neuronal cell body, the more likely the neuron is to die.
2. In the PNS, anterograde degeneration of axons is rapid and complete after several weeks.
3. In the PNS, the endoneurial Schwann cell sheath that envelops a degenerating axon does not degenerate and provides a scaffold for regeneration and remyelination of the axon.
4. In the CNS, anterograde degeneration of axons proceeds slowly and is complete only after several months.
5. In the CNS, oligodendrocytes withdraw their processes that form myelin from a degenerating axon and do not contribute to axon regeneration.
6. As degeneration and demyelination proceeds in the CNS, reactive astrocytes form a glial scar in place of the degenerating axon.

B. Neuron cell bodies undergo **retrograde chromatolysis** in response to **axotomy** (see Figure 2–10 A and B).
1. Chromatolytic changes are accompanied by increases in RNA and synthesis of proteins in an attempt to regenerate the axon.
2. In the first few days after axotomy, the Nissl substance breaks up and becomes dispersed.
3. The nucleus moves from a central position in the cell to an eccentric position adjacent to the cell membrane.
4. The cell body may swell, and the dendrites may shrink.
5. Axon terminals in synapse with the affected neuron may withdraw from the cell body and dendrites.
6. The chromatolytic process results in the growth of multiple sprouts from the proximal part of the cut axon.

C. **Neurons with severed axons** in the PNS are capable of complete axonal regeneration (see Figure 2–10B).
1. Successful sprouts from the cut axon grow into and through endoneurial sheaths and are guided by Schwann cells back to their targets.
2. Regeneration proceeds at the rate of 1–2 mm/day, which corresponds to the rate of slow anterograde transport.
3. Once the regenerated axon reaches the target, Schwann cells begin myelin production.
4. The diameter of the regenerated myelinated axon is usually smaller than that of the original axon, therefore, the conduction velocity of the nerve impulse is slower.
5. Axons in the PNS may not regenerate completely or may grow back to the wrong target (synkinesis).

NEUROMAS

- A **neuroma, a tangled mass of axonal sprouts,** may form at the site of a cut axon that fails to regenerate. Neuromas may cause extreme pain that is resistant to medication.
- Neuromas are associated with phantom limb pain that may arise after an amputation.

D. **Severed axons in the CNS** sprout small branches from the cut axon, but the axon fails to regenerate.

E. An **axon that is compressed** as a result of an irritative lesion in the CNS or PNS may also elicit retrograde changes in the cell bodies of affected neurons.
   1. There is no wallerian degeneration, and recovery occurs much more quickly.
   2. Axons may be compressed by a tumor, enlargement of a ventricle in patients with hydrocephalus, a mass effect herniation, or a hemorrhagic mass.

CLINICAL PROBLEMS

1. Your patient has succumbed to health problems related to Alzheimer's disease. An autopsy is performed. What might you expect to find in degenerating CNS neurons?

   A. Negri bodies

   B. Evidence of glutamate-induced excitotoxicity

   C. Retrograde chromatolysis

   D. Plaques near the surfaces of the lateral and fourth ventricles

   E. Neurofibrillary tangles

2. Your patient develops continuous contractions and spasms of axial and limb muscles and spasms of muscles of mastication and facial expression. What do these signs suggest that the patient has been infected with?

   A. Rabies

   B. Herpes

   C. Polio

   D. Neurosyphilis

   E. Tetanus

3. In a surgical procedure, a derivative of curare is used to induce muscle paralysis. By what mechanism does curare induce the paralysis?

   A. By binding to presynaptic calcium ion channels

   B. By binding to nicotinic receptors

   C. By inducing antibodies to bind to nicotinic receptors

   D. By binding to the presynaptic membrane

   E. By reducing the amount of ACh synthesized in the synaptic terminal

   F. By binding to acetylcholinesterase

4. The ventral root of the T1 spinal nerve has been cut.

   Which of the following might be expected?

   A. Retrograde chromatolysis in dorsal root ganglia of T1

   B. Wallerian degeneration between the site of the cut fibers and their neuron cell bodies

   C. Wallerian degeneration of axons distal to the cut

    D. Limited axonal regeneration because ventral roots are myelinated by oligodendrocytes

    E. Regeneration of severed axons at the rate of fast anterograde axonal transport

5. You are developing a drug to cross the blood–brain barrier.

    Which of the following is the most important element that maintains the integrity of the blood–brain barrier?

    A. Resistance to lipid-soluble substances

    B. Tight junctions between astrocyte foot processes

    C. Tight junctions between capillary endothelial cells

    D. Gap junctions between capillary endothelial cells

    E. Pericytes

6. A 25-year-old secretary comes to your office complaining of visual problems. Your examination reveals that she has blurry vision but no ptosis. She complains of tingling and numbness when she writes. When she walks, her gait is unsteady. Her stretch reflexes are hyperactive in the left lower limb but normal in the right lower limb. Cranial nerve reflexes and functions are normal except for her visual problem. A few months later, her neurological exam returns to normal except for some double vision and blurriness. What do the signs and symptoms suggest?

    A. Multiple sclerosis

    B. Guillain-Barré syndrome

    C. Myasthenia gravis

    D. Lambert-Eaton syndrome

    E. Amyotrophic lateral sclerosis

7. A 37-year-old woman came to her physician complaining of fatigue at work. At the end of day, she could barely hold her head up. She complained of not being able to speak clearly and had difficulty smiling in the evening and had droopy eyelids. When exercising, she found that she was breathing hard at the beginning of the workout, despite being in good physical shape. An examination revealed normal muscle tone, and her sensory exam for both cranial and spinal nerves was normal. However, after she was asked to gaze upward for 1 min, she could no longer keep her eyes open. What might the patient have?

    A. Lambert-Eaton syndrome

    B. Multiple sclerosis

    C. Myasthenia gravis

    D. Tetanus

    E. Guillain-Barré syndrome

8. Which of the following neural structures consists of axons that have the capacity to regenerate if cut?

    A. A tract

    B. A fasciculus

C. A dorsal root

    D. A dorsal column

    E. A lemniscus

9. A middle-aged male patient is a chronic alcoholic who complains of memory loss. The physician explains to the patient that ethanol may cause degeneration of neurons in many parts of the CNS that affect his ability to consolidate learned information into memories. How does ethanol cross the blood–brain barrier to reach the affected neurons?

    A. By retrograde axonal transport

    B. By receptor-mediated transport

    C. By diffusion

    D. By a mechanism facilitated by an ATPase motor protein

    E. Through ion channels

10. You are working for a pharmaceutical firm that is pursuing research on drugs that enhance long-term potentiation in neurons.

    At which of the following receptors might the drug act as an agonist to enhance LTP?

    A. A metabotropic receptor

    B. An NMDA receptor

    C. A D_2 receptor

    D. An AMPA receptor

    E. A G-protein coupled receptor

11. After a brief respiratory infection, a 55-year-old man is taken to the emergency room because he could not walk to the bathroom, and he complains that his legs are numb. A neurological exam reveals an absence of patellar and Achilles tendon reflexes bilaterally and a slight decrease in vibratory sense and in pain and temperature sensations in the lower limbs. His mental status and cranial nerve functions are normal. He seems to have trouble breathing in the hospital and is placed on a ventilator.

    What do the signs and symptoms suggest that the patient has?

    A. Multiple sclerosis

    B. Guillain-Barré syndrome

    C. Myasthenia gravis

    D. Lambert-Eaton syndrome

    E. Tetanus

12. A tritiated amino acid is injected into the vicinity of a neuron. The amino acid is taken up by the neuron, incorporated into a peptide neurotransmitter protein, and transported down the length of the axon. How fast is this protein transported down the length of the axon?

    A. 1–2 mm per day

    B. 5–10 mm per day

    C. 10–40 mm per day

D. 60–100 mm per day

    E. 100–400 mm per day

13. An anticancer drug acts by disrupting microtubules in cancer cells but also disrupts fast anterograde axonal transport. Movement of which of the following neurotransmitters or precursors of neurotransmitters will be most affected?

    A. Neuropeptides

    B. Glutamate

    C. Small molecule neurotransmitters

    D. Acetylcholine

    E. Amino acid neurotransmitters

14. Horseradish peroxidase and tritiated leucine are simultaneously injected into a synapse and adjacent to a neuron cell body, respectively. Assuming that both substances are incorporated at the same time into the cell, which of the following should be evident?

    A. The HRP will reach the cell body before the tritiated leucine reaches the axon terminal.

    B. The HRP will reach the cell body and the tritiated leucine will reach the axon terminal simultaneously.

    C. The tritiated leucine will reach the axon terminal before the HRP will reach the cell body.

    D. Dynein will facilitate anterograde transport of the tritiated leucine.

    E. Kinesin will facilitate retrograde transport of the HRP.

MATCHING PROBLEMS

Questions 15–21: Neurohistology match

Choices (each choice may be used once, more than once, or not at all):

    A. Myelin stain

    B. Nissl stain

    C. Golgi stain

    D. Anterograde transport of radiolabeled amino acids

    E. Retrograde transport of horseradish peroxidase

    F. None of the above

15. Used to demonstrate a neuron cell body undergoing retrograde chromatolysis.

16. Utilized to demonstrate sites of termination of axons.

17. Used to study shapes of all neuron cell bodies in a nucleus.

18. Demonstrates shape of cell body, branching pattern of dendrites, and course of axon.

19. Used to study the structure of synaptic vesicles at a synapse.

20. Used in studies to determine locations of neuron cell bodies that innervate a specific skeletal muscle.

21. Stains fibers in tracts and fasciculi in the spinal cord.

Questions 22–29: Neurotransmitter match. Match the neurotransmitter in choices A–E that correspond to the statement.

Choices: (each choice may be used once, more than once, or not at all):

    A. GABA

    B. Glutamate

    C. Acetylcholine

    D. Dopamine

    E. Norepinephrine

    F. More than one choice in A–E is correct

22. An inhibitory neurotransmitter used by neurons found throughout the CNS.

23. Used by preganglionic autonomic axons.

24. Used by autonomic neurons derived from neural crest cells.

25. Released by lower motor neurons.

26. An excitatory neurotransmitter used by 50% of CNS neurons.

27. Used by neurons in the nucleus accumbens.

28. Produced by neurons in the locus ceruleus.

29. Used by neurons that promote memory consolidation.

ANSWERS

1. The answer is E. Negri bodies are seen in patients with rabies, excitotoxicity and chromatolysis might be seen in stroke patients, and nonsenile plaques are a hallmark of a patient with MS.

2. The answer is E. Patients with tetanus have continuous contractions and spasms of axial and limb muscles and spasms of muscles innervated by certain cranial nerves. Patients with polio and rabies may have weakness of skeletal muscles innervated by the affected neurons. Herpes is taken up and retrogradely transported in sensory fibers and, when activated, causes painful vesicles in cutaneous areas. Neurosyphilis primarily affects the dorsal columns.

3. The answer is B. Curare binds to the nicotinic cholinergic receptors and blocks neuromuscular transmission.

4. The answer is C. Ventral roots do not contain sensory fibers, and retrograde changes occur between the site of the cut fibers and their neuron cell bodies. The roots may regenerate because their myelin is made by Schwann cells, but they regenerate at the rate of slow anterograde transport.

5. The answer is C. The blood–brain barrier permits lipid-soluble substances to readily diffuse across it. Astrocytes induce the capillary endothelial cells to form the

blood–brain barrier and maintain the tight junctions between them. Pericytes control proliferation and repair of endothelial cells, regulate vessel tone, and secrete elements of the basal lamina. Gap junctions do not link capillary endothelial cells.

6. The answer is A. In patients with MS, multiple lesions appear over time, but the signs and symptoms may undergo exacerbation and remission; commonly, 2 or more CNS sensory or motor neural systems are affected in separate attacks. MS is the most common cause of optic neuritis.

7. The answer is C. Patients with MG have weakness mainly in skeletal muscles innervated by cranial nerves. Ninety percent of diagnosed patients have ocular weakness and a bilateral ptosis.

8. The answer is C. All of the other choices are found inside the CNS and lack the capacity to regenerate as a result of myelination by oligodendrocytes.

9. The answer is C. Ethanol is a lipid-soluble substance that readily diffuses across the blood–brain barrier.

10. The answer is B. An NMDA receptor is the receptor site involved in LTP.

11. The answer is B. In Guillain-Barré syndrome, myelin formed by Schwann cells in the PNS undergoes an acute inflammatory reaction after a respiratory or gastrointestinal illness. Motor axons are always affected, producing weakness in the limbs. Weakness of cranial nerve innervated muscles or respiratory muscles may be seen.

12. The answer is A. Regeneration proceeds at the rate of slow anterograde axonal transport.

13. The answer is A. Microtubule mediated fast axonal transport delivers precursors of neuropeptides to the axon terminal

14. The answer is C. Anterograde axonal transport is faster than retrograde axonal transport

15. B

16. D

17. B

18. C

19. F

20. E

21. A

22. A

23. C

24. F (both C and E apply)

25. C

26. B

27. D

28. E

29. C

CHAPTER 3
SPINAL CORD

I. **The spinal cord is the simplest part of the central nervous system (CNS).**

  A. The spinal cord **extends from the foramen magnum approximately to the level of the body of the L2 vertebra** in adults.

  B. The spinal cord contains an **inner core of gray matter** that is completely surrounded by white matter (Figure 3–1).

   1. The gray matter is divided into a dorsal horn and a ventral horn, which are separated by an intermediate zone; the gray matter of the spinal cord has a "butterfly" shape in transverse sections.

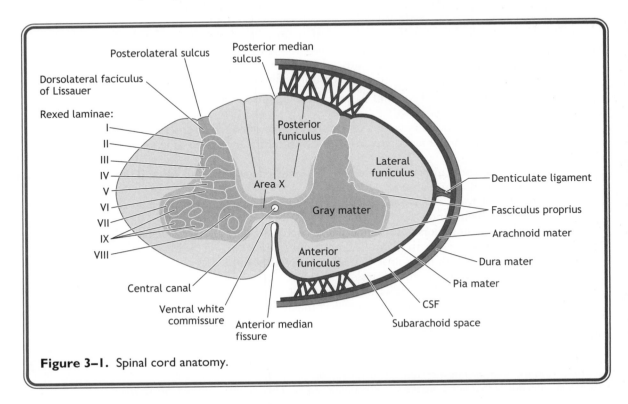

Figure 3–1. Spinal cord anatomy.

2. The gray matter of the spinal cord is organized into 10 Rexed laminae.
3. The dorsal horn consists of Rexed laminae I–VI and contains neurons derived from the alar plate that respond to sensory inputs provided mainly by dorsal roots of spinal nerves.
4. The ventral horn consists of Rexed laminae VIII and IX and contains the cell bodies of alpha and gamma motor neurons that innervate skeletal muscle. Lamina X surrounds the central canal.
5. The intermediate zone of gray matter is Rexed lamina VII, which contains preganglionic autonomic neuron cell bodies and the neuron cell bodies in Clarke's nucleus.
   a. Preganglionic sympathetic neurons in lamina VII are in the intermediolateral cell column; this column forms the lateral horn seen in transverse sections of the cord from T1 through L2.
   b. Axons from Clarke's nucleus ascend the spinal cord in the dorsal spinocerebellar tract and provide proprioceptive input to the cerebellum from the lower limbs.
6. Most Rexed laminae contain populations of interneurons that contribute to sensory processing or to voluntary and reflex contractions of skeletal muscles.
7. The white matter of the spinal cord is organized into funiculi, which contain tracts or fasciculi of axons that form components of sensory or motor neural systems (see Figure 3–1).
   a. Ascending tracts and fasciculi carry information to higher levels of the CNS.
   b. Descending tracts influence the voluntary or reflex contractions of skeletal muscles or influence preganglionic autonomic neurons.
   c. The white matter also contains intersegmental propriospinal axons that activate multiple segments of the cord in reflexes.
C. The **dorsal and ventral roots** of 31 pairs of spinal nerves enter or exit segmentally from the spinal cord.
   1. At the dorsal root entry zone, central processes of dorsal roots enter in a medial or lateral division and are classified using 2 different criteria (Figure 3–2, Table 3–1).
      a. **Dorsal roots of muscular nerves are classified according to fiber diameter** by Roman numerals.
      b. **Dorsal roots of cutaneous nerves are classified alphabetically** according to their conduction velocity.
   2. Medial division dorsal root fibers are large-diameter, heavily myelinated axons classified as Ia, Ib, II, and A-beta fibers (see Figure 3–2, Table 3–1).
      a. Class Ia and Ib fibers arise from muscle spindles and Golgi tendon organs (GTOs), receptors found in skeletal muscle and tendons, respectively.
      b. Class II fibers also innervate muscle spindles.
      c. A-beta fibers innervate mechanoreceptors in skin.

MEDIAL DIVISION DORSAL ROOT FIBERS

- *Medial division dorsal root fibers* are the fastest-conducting, largest-diameter afferent fibers. They are most sensitive to anoxia and most resistant to anesthesia.
- Lateral division dorsal root fibers are the slowest-conducting, smallest-diameter afferent fibers. They are more resistant than medial division fibers to anoxia and most sensitive to anesthesia.

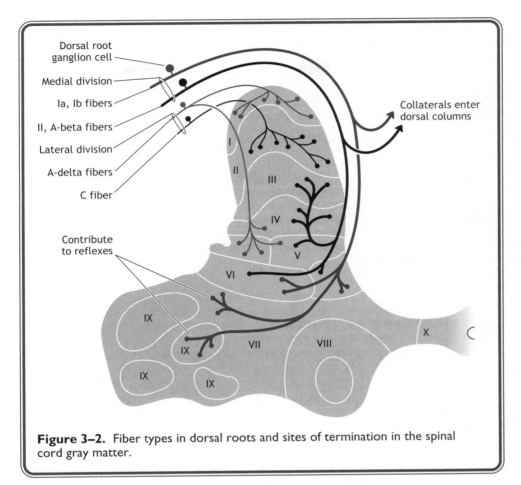

Figure 3–2. Fiber types in dorsal roots and sites of termination in the spinal cord gray matter.

3. Lateral division fibers are small-diameter, thinly myelinated (A-delta) or unmyelinated (C) fibers (see Figure 3–2, Table 3–1).
   a. A-delta fibers have free nerve endings that respond to the modalities of sharp pain and cold.
   b. C fibers have free nerve endings that respond to dull pain and warmth.
4. Three types of motor axons may exit the spinal cord in a ventral root (Figure 3–3, Table 3–2).
   a. Axons of alpha and gamma motor neurons exit in ventral roots from all spinal cord segments and innervate extrafusal skeletal muscle fibers at motor end plates (neuromuscular junctions) and intrafusal fibers in skeletal muscle (see later discussion).
   b. Preganglionic autonomic axons exit the spinal cord in the ventral roots from the T1 through L2 segments and from the S2 through S4 segments.

Table 3–1. Spinal nerves: fiber types in dorsal and ventral roots.

Fiber Type	Description
Classification Scheme	
I, A-alpha	Fastest-conducting, largest-diameter, most sensitive to anoxia, most resistant to anesthesia
II, A-beta	
III, A-delta	
IV, C	Slowest-conducting, smallest-diameter, least sensitive to anoxia, most sensitive to anesthesia

Fiber Type	Diameter (Microns)	Conduction Velocity (M/S)	Receptor	Function
Muscle Sensory Fibers[a]				
Ia	12–20	70–120	Muscle spindle afferent	Change in length or rate of change in length
Ib	12–20	70–120	Golgi tendon organ	Muscle force
II	6–12	30–70	Flower spray endings Joint receptors	Event of muscle stretch Joint angle
Fine Tactile (Epicritic) Sensory Fibers[a]				
A-beta	6—12	30—70	Merkel cell Meissner corpuscle Ruffini corpuscle Pacinian corpuscle Hair receptors Joint receptors	Pressure, texture Touch velocity Flutter (low-freq. vibration) Skin stretch Vibration Hair stroking Joint angle
Pain, Temperature (Protopathic), and Visceral Sensory Fibers[b]				
A-delta	1–5	5–30	Bare nerve endings	Sharp pain, cold, visceral receptors
C	0.2–1.5 (Unmyel.)	0.5–2	Bare nerve endings	Burning pain, warmth

Roman numerals I–IV designate fiber diameter (used for muscular nerves).
Letters A, B, C designate conduction velocity (used for cutaneous nerves).
[a]Enter spinal cord in medial division of dorsal root.
[b]Enter spinal cord in lateral division of dorsal root.

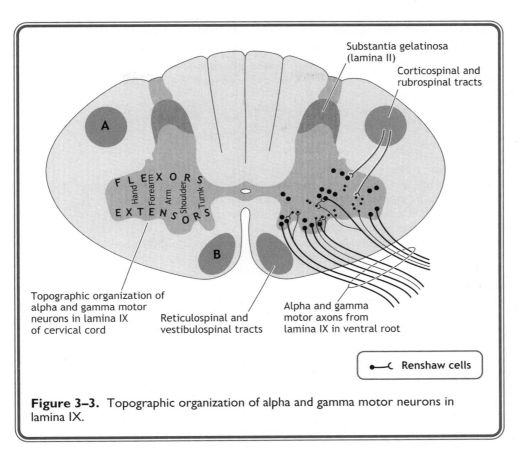

Figure 3–3. Topographic organization of alpha and gamma motor neurons in lamina IX.

Table 3–2. Motor fiber types in ventral root.

Fiber Type	Diameter (Microns)	Conduction Velocity (M/S)	Structure Innervated	Function
Alpha	13–20	70–120	Extrafusal skeletal muscle fibers	Contracts a motor unit
Gamma	2–10	15–40	Muscle spindles (intrafusal fibers)	Increases sensitivity of spindles to rate of stretch and change in muscle length
Preganglionic autonomic	1–3	3–15	Postganglionic autonomic neuron cell bodies	
Postganglionic autonomic	0.2–2	0.5–2	Smooth muscle autonomic glands, cardiac muscle	

II. Motor neural systems in the spinal cord provide voluntary and reflex control of skeletal muscles innervated by spinal nerves.

A. **Upper motor neurons** (UMNs) and **lower motor neurons** (LMNs) in the spinal cord provide a 2-neuron system for the voluntary control of skeletal muscles innervated by spinal nerves.

1. **There are 2 kinds of LMNs** with cell bodies in lamina IX of the ventral horn of the spinal cord (see Figures 3–2 and 3–3 and Table 3–2).

   a. Alpha motor neurons innervate extrafusal fibers in skeletal muscle at neuromuscular junctions.

   b. Gamma motor neurons innervate intrafusal fibers (muscle spindles) in skeletal muscle.

   c. The cell bodies of the alpha and gamma LMNs in the spinal cord are always on the same (ipsilateral) side of the midline of the CNS as the skeletal muscles that their axons innervate.

   d. LMNs in lamina IX that innervate axial muscles are medial to motor neurons that innervate muscles in the limbs.

   e. LMNs in lamina IX that innervate flexor muscles are dorsal to motor neurons that innervate extensor muscles.

2. **Upper motor neurons innervate lower motor neurons.**

   a. The cell bodies of UMNs are in the cerebral cortex and in the brainstem (Figure 3–4).

   b. The axons of UMNs descend into the spinal cord in tracts to reach and synapse directly with LMNs, or with interneurons which then synapse with LMNs.

   c. In most cases, the neuronal cell bodies of UMNs are on the opposite (contralateral) side of the midline of the CNS than the LMNs that they innervate.

   d. Structures in the brainstem that contain UMNs include the red nucleus, the pontine and medullary reticular formation, and the lateral and medial vestibular nuclei.

3. **The corticospinal, rubrospinal, and medullary (lateral) reticulospinal tracts arise from lateral or flexor-biased UMNs** (Figures 3–4 and 3–5).

   a. The largest and most important UMN tract for voluntary control of flexor muscles is the corticospinal tract.

   b. Most (60%) of the cell bodies of corticospinal tract axons are in the primary motor cortex, in the precentral gyrus of the frontal lobe, and the premotor area, located immediately anterior to primary motor cortex.

   c. Primary and secondary somatosensory cortical areas located in the parietal lobe give rise to about 40% of the fibers of the corticospinal tract.

   d. Axons in the corticospinal tract leave the cerebral cortex in the internal capsule and then descend through the ventral part of the midbrain, pons, and medulla.

   e. In the lower medulla, 80–90% of corticospinal fibers cross at the decussation of the pyramids; the crossed axons of the corticospinal tract descend the full length of the cord in the lateral funiculus of the white matter.

   f. The rubrospinal tract arises from the red nucleus in the midbrain; its axons cross just below the red nucleus and course adjacent to the corticospinal tract into the spinal cord.

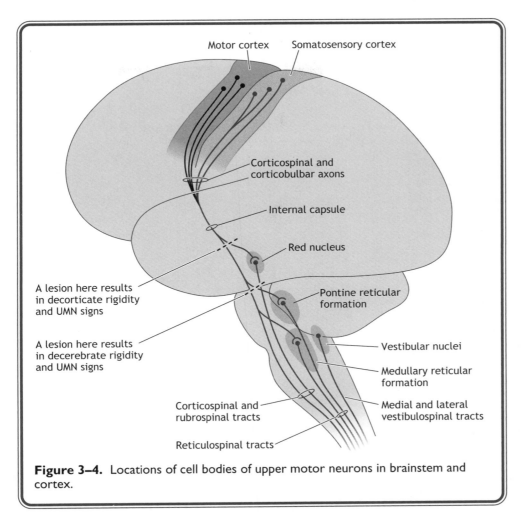

Figure 3–4. Locations of cell bodies of upper motor neurons in brainstem and cortex.

   g. The corticospinal tract activates alpha motor neurons that innervate flexor muscles that control skilled movements of the distal musculature in the limbs.

   h. The rubrospinal tract activates alpha motor neurons, which mainly innervate flexor muscles in the upper limb.

   i. The medullary (lateral) reticulospinal tract arises from neurons in the medullary reticular formation and, through interneurons, mainly inhibits alpha and gamma motor neurons that innervate flexor muscles.

4. The pontine reticulospinal and vestibulospinal tracts arise from extensor-biased UMNs (see Figures 3–4 and 3–5).

   a. The pontine (medial) reticulospinal tract arises from neurons in the pontine reticular formation and, through interneurons, activates motor neurons that innervate extensor muscles for postural support.

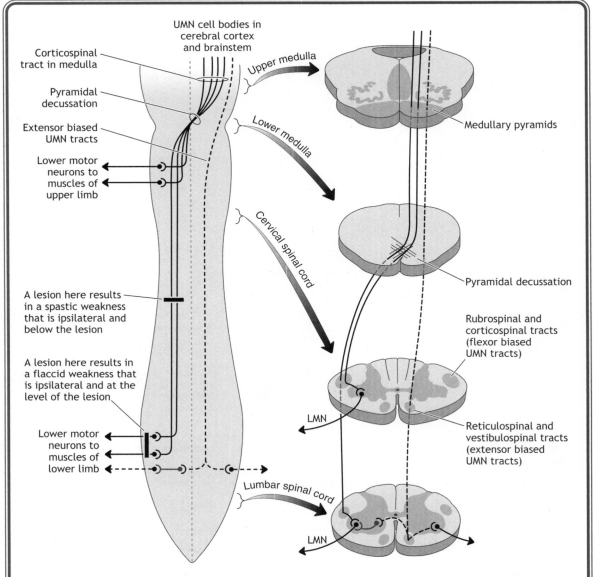

Figure 3–5. The course of axons of upper motor neurons in the medulla and spinal cord with representative cross-sections. Note the different effects of upper versus lower motor neuron lesions. (Reproduced with permission from Waxman, SG. *Clinical Neuroanatomy*, McGraw-Hill, 2003.)

b. The lateral vestibulospinal tract facilitates alpha motor neurons that innervate extensor muscles.

    c. The medial vestibulospinal tract facilitates alpha motor neurons that innervate neck and upper limb muscles in response to vestibular stimuli.

B. Reflex contractions of skeletal muscles use sensory receptors in skeletal muscle or skin, dorsal root fibers, interneurons, and lower motor neurons.

    1. The sensory neuron in a muscle reflex may be a proprioceptive afferent that is responding to a change in the stretch or force of a skeletal muscle, or it may be a cutaneous afferent responding to a noxious stimulus.

    2. The motor neuron in the reflex is the LMN. Contractions of skeletal muscles generated by reflexes do not require upper motor neuron input.

    3. The muscle stretch reflex uses muscle spindles, class Ia and class II muscle spindle dorsal root fibers, and lower motor neurons (Figures 3–6 and 3–7).

        a. The muscle stretch reflex provides the basis for muscle tone, the slight resistance to stretch found in all normally innervated skeletal muscles.

        b. Muscle spindles are intrafusal fibers in skeletal muscle that lie in parallel with the extrafusal fibers and are active when the muscle is lengthened or stretched.

        c. Stretching of the spindles stimulates the class Ia and II afferent fibers, which innervate the central region of the spindle, and the firing rate of these fibers increases in a manner that is directly proportional to the degree of stretch of the spindle.

        d. Class Ia afferent fibers monitor the length and rate of change in length of intrafusal fibers in response to changes in stretch of extrafusal fibers in skeletal muscle.

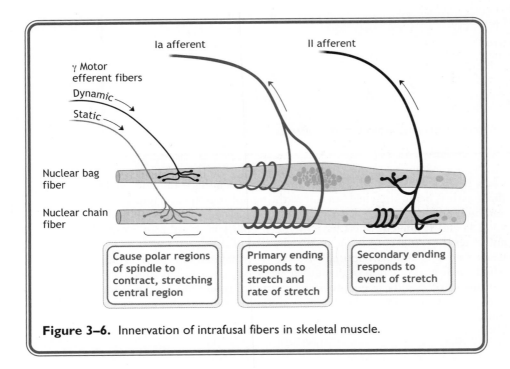

Figure 3–6. Innervation of intrafusal fibers in skeletal muscle.

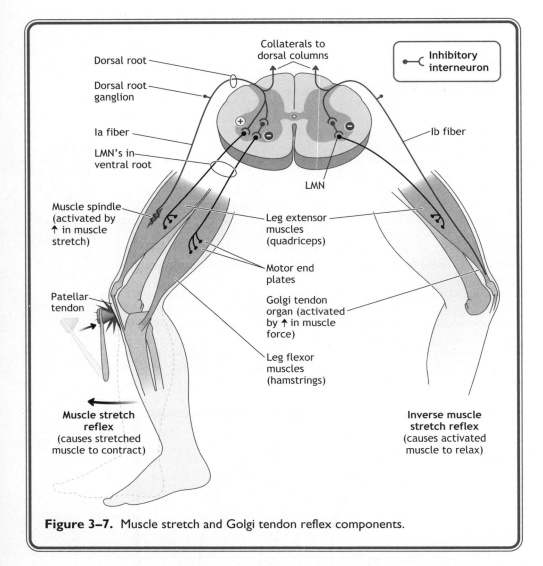

Figure 3–7. Muscle stretch and Golgi tendon reflex components.

e. Class II afferent fibers, which form flower spray endings, respond while the muscle is being stretched but not to the rate of stretch (see Figures 3–6 and 3–7).

f. Class Ia and II afferent fibers have their cell bodies in dorsal root ganglia and enter the spinal cord in the medial division of the dorsal root.

g. Class Ia fibers monosynaptically excite lower (alpha) motor neurons, which cause a contraction of the muscle being stretched and its agonists.

h. Collaterals of class Ia fibers provide a reciprocal inhibition by synapsing with interneurons that inhibit alpha motor neurons to the antagonist muscles.

IMPORTANT MUSCLES AND SPINAL CORD SEGMENTS TESTED IN MUSCLE STRETCH REFLEXES

- *In the upper limb, the **biceps brachii muscle stretch reflex** uses the C5 and C6 spinal cord segments, and the triceps brachii muscle stretch reflex uses the C7 and C8 spinal cord segments.*
- *In the lower limb, the **quadriceps femoris (patellar or knee jerk) reflex** uses the L3 and L4 spinal cord segments, and the gastrocnemius/soleus (Achilles tendon or ankle jerk) reflex uses the S1 and S2 spinal cord segments.*

CLINICAL GRADING OF MUSCLE STRETCH REFLEXES

***Muscle stretch reflexes are graded from 0 (absence of muscle stretch reflexes) to 4 (hyperactive stretch reflexes with clonus)**. A grade of 2+ indicates normal, moderately brisk reflexes.*

CLINICAL GRADING OF MUSCLE STRENGTH

*The **grading of muscle strength** ranges from 0/5 (patient is unable to generate contractions that produce movement) to 5/5 (normal strength; patient generates movement against full resistance).*

4. **The Golgi tendon reflex** uses Golgi tendon organs (GTOs), class Ib dorsal root sensory fibers, interneurons, and alpha motor neurons.
   a. GTOs are found in tendons of skeletal muscles and respond to changes in muscle force or tension.
   b. Class Ib dorsal root afferent fibers innervate GTOs, have their cell bodies in dorsal root ganglia, and enter the spinal cord in the medial division of the dorsal root.
   c. The Ib afferents enter and synapse with a variety of interneurons; the interneurons regulate the activity of alpha motor neurons by distributing the force generated by the contracting muscle by inhibiting or activating alpha motor neurons to that muscle (an autogenic inhibition).
   d. If excessive force builds up in a skeletal muscle, the Ib fibers activate inhibitory interneurons, which inhibit alpha motor neurons to the muscle generating the excessive force, causing that muscle to relax.
   e. Collaterals of Ib fibers provide a reciprocal excitation by stimulating interneurons that activate alpha motor neurons to antagonist muscles.

THE CLASP KNIFE REFLEX AND GOLGI TENDON REFLEX

*The **clasp knife reflex** is activated when a muscle is stretched during an isometric contraction. The attempted stretch causes a rapid buildup of force in the muscle, a large discharge by the GTOs and Ib afferents, and a sudden inhibition of the affected muscle. Weight lifters may exhibit this reflex when they attempt to lift a particularly heavy load. A hyperactive clasp knife reflex is also seen in patients with an upper motor neuron lesion (see later discussion).*

5. The gamma loop facilitates contraction of skeletal muscles by using gamma motor neurons, Ia afferent fibers, interneurons, and alpha motor neurons.
   a. Gamma motor neurons provide an efferent innervation to muscle spindles but not to Golgi tendon organs (GTOs) (see Figure 3–7).
   b. Axons of gamma motor neurons cause the polar regions of the muscle spindle to contract; the central region of the spindle is stretched, which

results in an increase in the firing of Ia fibers that innervate the central region of the spindle.

   c. Ia fibers enter the spinal cord in the dorsal root and form monosynaptic excitatory synapses directly with the alpha motor neurons that innervate the same muscle, facilitating contraction of the extrafusal fibers of that muscle.

  6. Alpha-gamma coactivation keeps muscle spindles active, or "on the air," during muscle contractions.

   a. UMN systems stimulate both gamma and alpha motor neurons when a voluntary contraction of extrafusal fibers in a skeletal muscle is desired (see Figure 3–3).

   b. Activation of alpha motor neurons causes the extrafusal fibers to contract, but shortening the muscle also shortens and "unloads the muscle spindles," causing the Ia fibers to decrease or stop their firing, reducing their proprioceptive input to the spinal cord.

   c. The gamma motor neurons offset the shortening of the muscle spindles by causing contraction of the polar regions of the spindles, which stretches the central region of the spindles (see Figure 3–6).

   d. The firing rate of the Ia fibers is maintained by the gamma innervation, and the spinal cord continues to receive proprioceptive input from the contracting muscle.

MUSCLE CONTRACTIONS AND ALPHA-GAMMA COACTIVATION

Alpha-gamma coactivation is important when muscles are contracting against an unexpected load. The muscle spindles are stretched by the load but respond by increasing the firing of alpha motor neurons through the gamma loop.

  7. Recurrent inhibition by Renshaw cells limits the firing of alpha motor neurons.

   a. Renshaw cells are inhibitory interneurons found in lamina IX of the spinal cord gray matter that use glycine as a neurotransmitter.

   b. Recurrent collaterals of axons of alpha motor neurons innervate Renshaw cells, through ACh binding to muscarinic receptors.

   c. Renshaw cells provide a recurrent, or feedback, inhibition by synapsing with the same alpha motor neurons.

   d. Renshaw cells provide a mechanism to adjust the sensitivity of pools of alpha motor neurons and inhibit the firing of alpha motor neurons after a brief period of excitation.

TETANUS TOXIN AND RELEASE OF GLYCINE FROM RENSHAW CELLS

In the spinal cord, **tetanus toxin** inhibits release of glycine from synaptic terminals of Renshaw cells. Patients may have sustained contractions and spasms of axial and limb muscles. If cranial nerves are involved, patients may have spasms of muscles of mastication (trismus, or lockjaw) and of muscles of facial expression.

  C. The **flexor withdrawal and crossed extensor reflexes** are protective reflexes that use cutaneous receptors and the dorsal root fibers that innervate them, interneurons, and lower motor neurons (Figure 3–8).

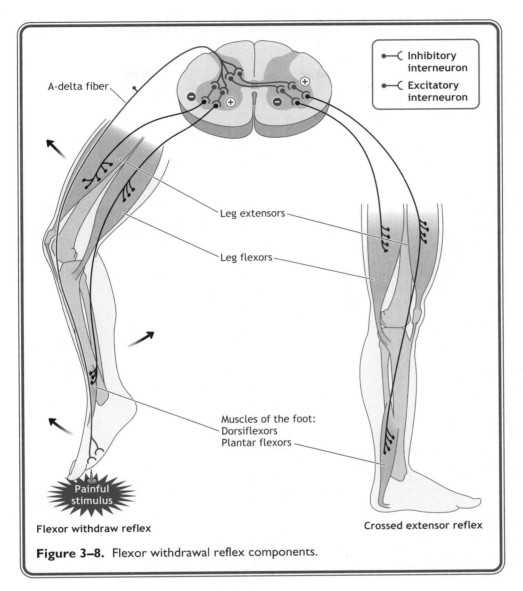

Figure 3–8. Flexor withdrawal reflex components.

1. These reflexes are activated by a noxious or excessive thermal stimulation of the skin that excites A-delta and class C dorsal root fibers.
2. These fibers enter the spinal cord in the lateral division of the dorsal root and activate multiple interneurons within the gray matter.
3. Through polysynaptic pathways, the interneurons facilitate LMNs to flexor and extensor muscles on both sides of the spinal cord.
4. Contractions of flexor muscles on the side of the stimulus facilitate removal of a limb from the painful stimulus, and contractions of extensors in the opposite limb provide postural support during limb withdrawal.

LESIONS OF LOWER MOTOR NEURONS RESULT IN A FLACCID PARALYSIS (FIGURE 3–9A)

*Patients with a **flaccid paralysis** may have the following:*

- *An absence of voluntary or reflex contractions of denervated muscles (paralysis).*
- *An absence of or suppressed muscle stretch reflexes (hypo- or areflexia, hypotonicity).*
- *Fasciculations, random twitches of denervated motor units visible beneath the skin.*
- *Atrophy or wasting of the denervated muscles.*
- *Fibrillations, which are invisible 1- to 5-ms potentials, detected with electromyography.*

*With few exceptions, LMN lesions result in a flaccid paralysis that is **ipsilateral to** and **at** the level of the lesion.*

POLIOMYELITIS, WERDNIG-HOFFMANN DISEASE, AND LMNS

- ***Poliomyelitis results from a loss of LMNs in the ventral horn caused by the poliovirus**. The disease causes a flaccid paralysis of muscles with the accompanying hyporeflexia, fasciculations, and atrophy.*
- *Several forms of **Werdnig-Hoffmann disease**, or infantile spinal muscular atrophies (SMAs), result in destruction of LMNs in infants or young children. Infants may have difficulty sucking, swallowing, or breathing and weakness in the limbs; they are "floppy" babies.*

UPPER MOTOR NEURON LESIONS RESULT IN A SPASTIC WEAKNESS OR PARESIS (SEE FIGURE 3–9B)

Patients may have the following:

- *Elevated or hyperactive muscle stretch reflexes (hyperreflexia or hypertonicity) because UMN systems have a net overall inhibitory effect on muscle stretch and inverse muscle stretch reflexes.*
  –The hyperactive muscle stretch reflexes seen in patients with UMN lesions is mainly due to a loss of inhibition of gamma motor neurons by reticulospinal medullary UMNs.
  *–Patients with hyperactive muscle stretch reflexes may have **clonus**, which is characterized by rapid successive reflex contractions and relaxations of agonists and antagonists observed mainly at the knee and ankle joints during reflex testing.*
- *A clasp knife reaction is an indication of a loss of upper motor inhibition of the inverse muscle stretch reflex.*
  –In a clasp knife reaction, passive stretch of a hypertonic muscle abnormally increases muscle force and increases the activity of Golgi tendon organ (GTOs).
  –The increased activity in GTOs results in a sudden release of resistance when the muscle is stretched and a relaxation of the hypertonic muscle.
- *Altered cutaneous reflexes. The Babinski sign is the best-known example of an altered cutaneous reflex.*
  –Normally, the toes flex in response to a plantar cutaneous stimulus. Patients with a UMN lesion may have a Babinski sign, characterized by an extensor plantar response (extension of the great toe and fanning of the other toes).
  –A Babinski sign is normal in infants until the UMN tracts become fully myelinated.
  –Two other cutaneous reflexes—the abdominal reflex, which uses the T8 through T12 spinal segments, and cremasteric reflex, which uses the L1 spinal segment—are paradoxically absent in patients with UMN lesions.
  –In a normal abdominal reflex, stroking the skin of the abdomen results in reflex contraction of abdominal wall muscles with movement of the umbilicus toward the stimulus.
  –In a normal cremasteric reflex, stroking the skin of the medial thigh in the male normally produces a visible contraction of the cremaster muscle and elevation of the testis.

A Poliomyelitis, Werdnig-Hoffman disease

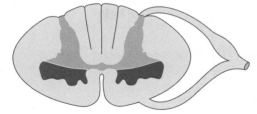

B UMN lesion (e.g., multiple sclerosis)

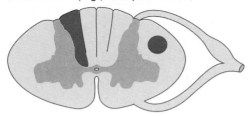

C Amyotrophic lateral sclerosis

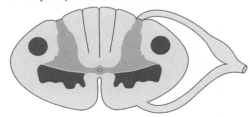

D Dorsal column lesion (e.g., multiple sclerosis)

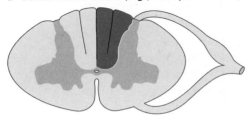

E Tabes dorsalis

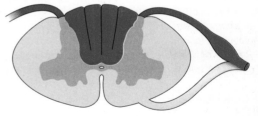

F Syringomyelia (early)

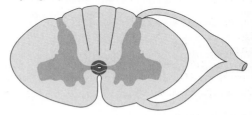

G Syringomyelia (late)

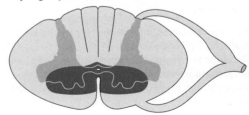

Figure 3–9. A–G: Structures affected in common spinal cord lesions I.

- *Atrophy of weakened muscles only as a result of disuse because these muscles can be made to contract by stimulating muscle stretch reflexes.*
  - *Lesions of UMNs result in a spastic paresis that may be* **ipsilateral** *or* **contralateral to** *the lesion and is always below the lesion.*
  - *A lesion of axons of UMNs in the spinal cord results in a spastic paresis that is ipsilateral and below the level of the lesion.*
  - *A lesion of UMNs between the cerebral cortex and the medulla* **above the decussation** *of the pyramids will result in a spastic paresis that is contralateral and below the level of the lesion.*

DECORTICATE VERSUS DECEREBRATE POSTURING IN PATIENTS WITH UMN LESIONS

- *Lesions of tracts of UMNs* *(e.g., corticospinal tract) above the red nucleus in the midbrain may result in decorticate posturing (see Figure 3–4). In these patients, there are contractions of muscles that act to flex at joints of the upper limb and contractions of muscles that act to extend at joints of the lower limb.*
- *Lesions of UMN systems in the brainstem below the red nucleus in the midbrain may result in decerebrate posturing. In these patients, there may be postural extension at joints of both the upper and lower limbs.*

AMYOTROPHIC LATERAL SCLEROSIS AND UPPER AND LOWER MOTOR NEURONS (FIGURE 3–9C)

- *Amyotrophic lateral sclerosis* *(ALS, or Lou Gehrig's disease) is a pure motor system disease that affects both UMNs and LMNs.*
- *ALS begins in the lower cervical segments of the cord that supply the upper limbs (40% of patients) or in the lumbosacral segments that supply the lower limbs (40% of patients). LMNs in cranial nerves are affected in 20% of patients.*
- *In ALS at lower cervical levels, patients have bilateral flaccid weakness at the level of the lesion in muscles of the upper limbs and bilateral spastic weakness below the lesion in muscles of the lower limbs. If ALS progresses superiorly from cervical levels, LMNs of the phrenic nerve will be affected, compromising respiration.*
- *Patients with ALS at caudal brainstem levels may have difficulty swallowing and difficulty speaking because of a weakness of pharyngeal or tongue muscles innervated by cranial nerves X and XII.*

III. **Descending hypothalamic axons and pontine upper motor neurons control preganglionic autonomic neurons.**

    A. The **descending hypothalamic axons** arise from cell bodies in the posterior hypothalamus and synapse with preganglionic sympathetic neurons in the intermediolateral cell column from the T1 through L2 spinal cord segments.

    B. **Axons of pontine upper motor neurons** and neurons in the anterior hypothalamus descend to sacral spinal levels to synapse with preganglionic parasympathetic neurons in the S2 through S4 segments and control the bladder.

LESIONS OF THE DESCENDING HYPOTHALAMIC AXONS, HORNER'S SYNDROME, AND ORTHOSTATIC HYPOTENSION

- *A lesion of the descending hypothalamic axons between the hypothalamus and the preganglionic sympathetic neurons in the T1 spinal segment may result in Horner's syndrome. Horner's syndrome is caused by disruption of a hypothalamic/sympathetic pathway, which provides sympathetic innerva-*

tion to the sweat glands in the face and scalp, blood vessels in the head, and the superior tarsal and dilator pupillae muscles in the orbit. Patients with Horner's syndrome have miosis (pupillary constriction), ptosis (drooping eyelid), and anhidrosis (lack of sweating), signs that are always ipsilateral to the side of the lesion.
- Patients may also have orthostatic hypotension in which there is a significant drop in systemic blood pressure when the patient changes from a lying to a standing position.
- A lesion of the bladder control axons from the pons above sacral levels S2 through S4 may result in a spastic bladder. In these patients, sacral parasympathetic neurons that innervate the bladder are not inhibited effectively when the bladder is stretched during filling. The bladder contracts in response to a minimum amount of stretch, causing frequent involuntary reflex emptying.

IV. **Two sensory systems, the dorsal column/medial lemniscal system and the anterolateral system, use 3 neurons to convey proprioceptive, epicritic, and protopathic sensations from receptors to the cerebral cortex.**

   A. In both systems, the **first, or primary, sensory neuron that innervates a sensory receptor** has a cell body in a dorsal root ganglion and enters the spinal cord in a dorsal root of a spinal nerve.

   1. The first neuron synapses with a second neuron in the brainstem or the spinal cord; the cell body of the second neuron is on the same side of the midline of the CNS as the entry point of the first neuron.
   2. The axon of the second neuron crosses the midline in the vicinity of the neuron cell body and ascends to the thalamus in a tract or a lemniscus.
   3. The axon of the second neuron synapses on a third neuron that is in a specific nucleus in the thalamus, and the axon of the third neuron projects to the primary somatosensory cortex.

   B. Each **sensory receptor transduces or converts the energy of a stimulus** into a generator potential, a graded change in the membrane voltage of the primary sensory neuron.

   1. If the generator potential is large enough, an action potential will be generated, and the impulse will be conducted by primary afferent neurons into the spinal cord.
   2. The intensity of the stimulus is coded by the frequency of the action potentials.
   3. As the stimulus strength is increased, more ion channels open, and the generator potential increases.
   4. Some receptors are rapidly adapting; in response to a sustained stimulus, the frequency of action potentials in the primary afferent neuron decreases even though the stimulus remains constant.
   5. Other receptors are slowly adapting; the frequency of action potentials in the primary afferent neuron decreases slowly in response to a sustained stimulus.

   C. All **receptors are defined by a receptive field**, the area of sensory space (e.g., skin) in which a stimulus influences a receptor.

V. **The dorsal column/medial lemniscal system uses 3 neurons to convey proprioceptive and epicritic sensations to the cerebral cortex from the neck, trunk, and limbs (Figure 3–10).**

   A. **Proprioceptive sensations** enable perception of the position of the limbs and body in space based on the degree of stretch or contraction of skeletal muscles.

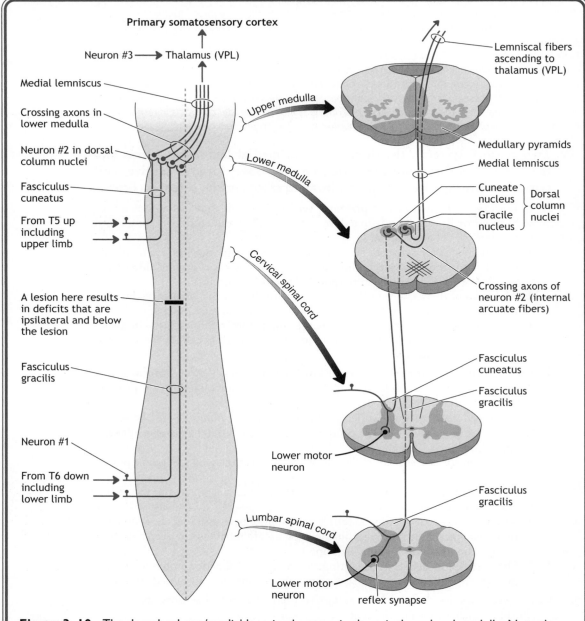

Figure 3–10. The dorsal column/medial lemniscal system in the spinal cord and medulla. Note the effect of a dorsal column lesion. (Reproduced with permission from Waxman, SG. *Clinical Neuroanatomy,* McGraw-Hill, 2003.)

B. Epicritic sensations include the discriminative touch modalities that are detected by a variety of encapsulated receptors and contribute to cutaneous perception of the size, shape, texture, and movement of objects (see Table 3–1).

C. The **primary afferent neurons** in this system have their cell bodies in the dorsal root ganglia and enter the cord via large-diameter, myelinated Ia, Ib, and class II or A-beta dorsal root fibers (see Table 3–1).

D. The **afferent fibers** enter in the medial division of the dorsal root and ascend the cord in the dorsal columns in the dorsal funiculus.

1. The **2 dorsal columns** are the **fasciculus gracilis** and the **fasciculus cuneatus**.
2. The fasciculus gracilis is found at all spinal cord levels, is situated close to the midline, and conveys sensations from skin and muscles in the lower limb and inferior aspect of the trunk beginning at and below the T6 dermatome.
3. The fasciculus cuneatus is found only at upper thoracic (above T5) and cervical spinal cord levels, is lateral to the fasciculus gracilis, and conveys sensations from skin and muscles in the upper trunk beginning at and above the T5 dermatome, including the upper trunk, upper limb, neck, and posterior scalp.

E. Axons in the dorsal columns ascend the length of the spinal cord to synapse with second neurons in the caudal medulla (see Figure 3–10).

1. The cell bodies of the second neuron in the caudal medulla are in the nucleus gracilis and nucleus cuneatus.
2. Cells in these medullary nuclei give rise to axons that cross the midline in the vicinity of their cell bodies (as internal arcuate fibers) and form the medial lemniscus.

F. Crossed axons in the medial lemniscus ascend through the length of the brainstem and project to the third neuron in the ventral posterolateral (VPL) nucleus of the thalamus.

G. Thalamocortical axons from the third neuron in the VPL nucleus project to primary somatosensory cortex in the postcentral gyrus, located in the anterior part of the parietal lobe.

LESIONS OF THE DORSAL COLUMNS AND AN EPICRITIC SENSORY LOSS

- A *lesion of the dorsal columns* results in a loss of epicritic sensations (see Figure 3–9D), including joint position sensation, vibratory and pressure sensations, and 2-point discrimination that is ipsilateral and begins immediately below the lesion. Patients with degenerative or demyelinating diseases may have paresthesias, altered sensations of touch and may not know where a limb is in space.
- Typically, the dorsal column/medial lemniscal system is evaluated by testing vibratory sense using a 128-Hz tuning fork.

TABES DORSALIS, DORSAL ROOT GANGLIA, AND AXONS IN THE DORSAL COLUMNS

- *Tabes dorsalis* is one manifestation of neurosyphilis and results in bilateral degeneration of the large-diameter dorsal root fibers and their neuron cell bodies in dorsal root ganglia and degeneration of the dorsal columns (most commonly the fasciculi gracilis), which contain their axons (see Figure 3–9E).
- Tabetic patients commonly have the 3 "Ps": **paresthesias, pain, and polyuria**.

- **Paresthesias** result from impaired vibration and position sense in the lower limbs carried by the fasciculus gracilis. The radiating **pain** results from hypersensitivity of the small-diameter A-delta and class C pain and temperature dorsal root fibers that are irritated in tabetic patients. **Polyuria** results from the loss of large-diameter sensory neurons that mediate bladder fullness resulting in the frequent emptying of the bladder. Urine retention may also be evident. Muscle stretch reflexes may be suppressed in the lower limbs as a result of degeneration of the Ia dorsal root fibers.
- Patients with tabes dorsalis may also have **Argyll Robertson pupils**, which accommodate, but are unreactive to, light.

THE ROMBERG TEST AND SENSORY VERSUS MOTOR ATAXIA

CLINICAL CORRELATION

- **The Romberg test** may be used to distinguish between a lesion of the dorsal columns, which results in sensory ataxia, and a lesion of the midline (vermal area) of the cerebellum, which results in motor ataxia.
- In the Romberg test, the patient stands with the feet together while steadied by the examiner and then is asked to close the eyes. A positive Romberg sign is elicited if the patient sways with eyes closed, indicating that the patient has a sensory ataxia. With the eyes open, visual input to the cerebellum helps the patient maintain balance despite the loss of proprioceptive input provided by the dorsal columns. If the patient has balance problems and sways with eyes open, this may indicate a lesion to the cerebellum and a motor ataxia.

VI. **The anterolateral system uses 3 neurons to convey protopathic sensations from the neck, trunk, and limbs (Figure 3–11).**

    A. **Protopathic sensations** include modalities such as nociception, which is the perception of pain or itch because of tissue damage or irritation, and temperature, which is the perception of warmth or cold.

    B. The **primary afferent neurons** have their cell bodies in the dorsal root ganglia and arise mainly from free nerve endings in the skin, subcutaneous tissues, muscles, and joints.

    C. **Dorsal root afferents** conveying pain and temperature enter the cord in the lateral division of the dorsal root by way of small-diameter, thinly myelinated or unmyelinated A-delta or class C dorsal root fibers (see Table 3–1).

        1. Thinly myelinated A-delta dorsal root fibers innervate temperature receptors, which are activated by extreme temperatures (> 45°C or < 5°C).

        2. A-delta dorsal root fibers also innervate mechanoreceptors, which are activated by intense pressure on the skin.

        3. Unmyelinated class C fibers innervate polymodal receptors, which are activated by high-intensity thermal or mechanical stimuli.

    D. In the skin, **tissue damage results in the release of prostaglandins and bradykinin**, activating pain- and temperature-sensitive free nerve endings of the primary afferent neurons, which act as receptors.

        1. Collaterals of these neurons release substance P and calcitonin gene-related peptide (CGRP) back into the adjacent epithelial cells.

        2. Substance P causes mast cells to release histamine, which stimulates the polymodal receptors.

        3. Substance P also promotes plasma extravasation, and CGRP dilates blood vessels; both actions promote further release of bradykinin.

        4. Bradykinin is a potent pain-producing agent that directly stimulates both A-delta and C fibers and promotes further release of prostaglandins.

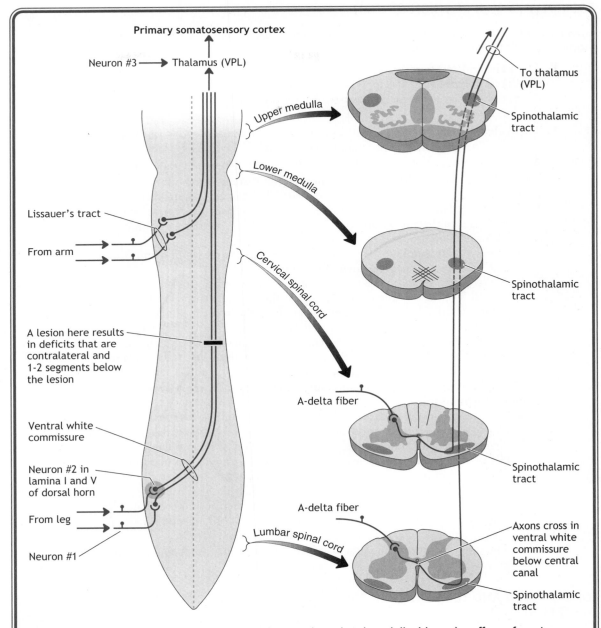

Figure 3–11. The anterolateral system in the spinal cord and medulla. Note the effect of a spino-thalamic tract lesion. (Reproduced with permission from Waxman, SG. *Clinical Neuroanatomy,* McGraw-Hill, 2003.)

ASPIRIN AND SUPPRESSION OF PAIN

Aspirin and other nonsteroidal analgesics are effective as pain suppressants because they block an enzyme used in the synthesis of prostaglandins.

    E. The **pain and temperature dorsal root fibers** are much shorter than those of the dorsal column/medial lemniscal system and have collaterals that ascend or descend 1-2 cord segments in the dorsolateral tract of Lissauer before entering the dorsal horn.

    F. The **A-delta fibers** synapse mainly with neurons in laminae I and V of the dorsal horn. The C fibers synapse mainly with lamina II neurons but also on the dendrites of lamina V neurons, which extend into lamina II.

    G. **Three second axon pathways in the anterolateral system** arise from neurons in the dorsal horn and cross in the vicinity of their cell bodies at all spinal cord levels.

       1. The spinothalamic tract is the main second axon pathway that carries modalities of pain and temperature and arises mainly from neurons in laminae I and V of the dorsal horn.

          a. Axons of laminae I and V neurons cross in the ventral white commissure just ventral to the central canal of the spinal cord.

          b. Axons of the second neurons in the spinothalamic tract cross at every segmental level of the spinal cord; axons of the second neuron in the dorsal column/medial lemniscal system cross only in the lower medulla.

          c. The spinothalamic tract axons coalesce in the ventral part of the lateral funiculus and course through the length of the spinal cord and the brainstem to terminate in the VPL nucleus of the thalamus. The thalamic neurons project to somatosensory cortex.

       2. **The spinoreticular tract and the spinomesencephalic tract also convey pain information**.

          a. The spinoreticular tract arises from neurons in laminae VII and VIII, is largely uncrossed, and terminates in the brainstem reticular formation.

          b. The spinoreticular tract provides pain information to the intralaminar nuclei of the thalamus, which are associated with arousal; the intralaminar nuclei project to the limbic system, which processes the emotional content of pain and aids in the memory of a stimulus that resulted in pain.

          c. The spinomesencephalic tract arises from neurons in laminae I and V, courses dorsally in the white matter of the spinal cord, and terminates in the periaqueductal gray region of the midbrain.

          d. The spinomesencephalic tract activates neurons in the periaqueductal gray, which suppress pain (see later discussion).

LOSS OF PAIN AND TEMPERATURE AND LESIONS OF ANTEROLATERAL SYSTEM STRUCTURES

- *Because most pain and temperature information crosses within 1–2 segments after it enters the spinal cord, a unilateral lesion of the crossed axons of the spinothalamic tract in the spinal cord or brainstem results in a loss of pain and temperature* **contralateral to and below the lesion.**
- *If the spinothalamic tract is lesioned in the spinal cord, the loss of pain and temperature is contralateral and begins 1–2 segments below the level of the lesion.*
- *If dorsal roots conveying pain and temperature are lesioned, there will be a loss of pain and temperature that is ipsilateral and limited to the dermatomal distribution of the affected roots at the level of the lesion.*

SYRINGOMYELIA AND THE CROSSING AXONS
OF THE SPINOTHALAMIC TRACTS

- *Syringomyelia* (Figure 3–12) results from a cavitation of the central canal of the spinal cord, which most commonly begins in the cervical spinal cord segments but may involve other cord regions or the medulla (see Figure3–9F and G). A syrinx is frequently associated with an Arnold-Chiari malformation where there is a downward herniation of the cerebellar vermis through the foramen magnum (Figure 3–12). Initially, patients have a bilateral loss of pain and temperature at or just below the level of the cavitation as a result of compression of spinothalamic axons crossing in the ventral white commissure just below the central canal. As the cavitation expands, lower motor neurons in 1 or both ventral horns may be compressed, resulting in flaccid paralysis of upper limb muscles innervated by the affected LMNs.
- Unilateral or bilateral Horner's syndrome may be a late manifestation of a cervical cavitation, which occurs as a result of compression of descending hypothalamic axons adjacent to the ventral horns. Infants born with an Arnold-Chiari malformation may also have syringomyelia.

  H. The **dorsal horn and brainstem** contain neurons that act to control pain (Figure 3–13).
  1. Within the dorsal horn, a gate control mechanism acts to suppress sensations of pain conveyed into the spinal cord by both the A-delta and class C dorsal root fibers that synapse either directly or indirectly with a spinothalamic projection neuron.

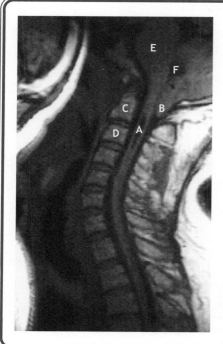

Figure 3–12. Sagittal MRI of the vertebral column showing a cervical cord syrinx (**A**) combined with Arnold-Chiari malformation (**B**). **C:** Odontoid process of the C2 vertebra. **D:** Body of C3 vertebra. **E:** Brainstem Note similarity of density of CSF in fourth ventricle (**F**) with that of syrinx.

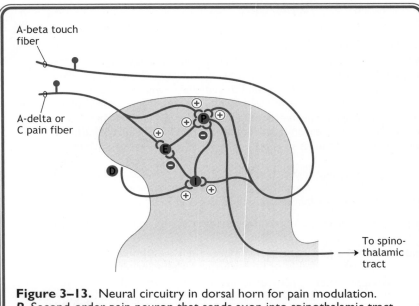

Figure 3–13. Neural circuitry in dorsal horn for pain modulation.
P. Second-order pain neuron that sends axon into spinothalamic tract.
E. Excitatory interneuron. **I.** Inhibitory interneuron. **D.** Axons that
descend from lateral tegmental and raphe nuclei to suppress pain (Re-
produced with permission from Michael-Titus, Revest, and Shortland.
The Nervous System, Elsevier, 2007.)

2. Pain may be suppressed by the interactions of the projection neuron with an
   incoming A-beta touch fiber or by descending input from the brainstem
   (Figure 3–13).
   a. Nociceptive information conveyed by the pain fibers opens a gate to pain by
      directly stimulating a spinothalamic projection neuron or by stimulating an
      excitatory interneuron, which in turn stimulates the projection neuron.
   b. Application of a cold, warm, or touch stimulus to the area of the painful
      stimulus stimulates A-beta nonnociceptive dorsal root fibers.
   c. The A-beta nonnociceptive dorsal root fibers counteract the pain fiber
      input and close the gate to pain by stimulating the projection neuron or
      by stimulating an inhibitory interneuron.
   d. The A-beta fiber also stimulates spinothalamic neurons to respond to a
      nonnoxious stimulus.
3. Serotonergic and noradrenergic neurons in the brainstem also act to suppress
   pain.
   a. Neurons with opioid receptors in the periaqueductal gray of the midbrain
      are activated by the spinomesencephalic tract and by the amygdala in the
      limbic system.
   b. Neurons in the periaqueductal gray project to serotonergic neurons, in the
      brainstem raphe nuclei, and to noradrenergic neurons in the lateral teg-
      mental nucleus.

c. Both the raphe and the lateral tegmental nucleus axons descend into the spinal cord to synapse on inhibitory (enkephalinergic) interneurons in the substantia gelatinosa (lamina II).

VII. The spinocerebellar system uses 2 neurons to convey unconscious proprioceptive input from muscle spindles and Golgi tendon organs (GTOs) to the cerebellum (Figure 3–15).

A. The **cerebellum uses proprioceptive input** to coordinate smooth execution generated by contractions of skeletal muscles.

B. **Proprioceptive input** is conveyed into the spinal cord by way of Ia and Ib dorsal root fibers, which synapse with second neurons in Clarke's nucleus in lamina VII or in the external cuneate nucleus in the caudal medulla.

C. The **dorsal spinocerebellar tract** arises from Clarke's nucleus from T1–L2 and carries proprioceptive input from muscles and tendons in the ipsilateral lower limb and lower trunk.

D. The **cuneocerebellar tract** (CCT) arises from the external cuneate nucleus and carries proprioceptive input to the cerebellum from muscles and tendons in the ipsilateral upper limb and upper trunk.

E. The **ventral spinocerebellar tract** (VSCT) arises from spinal border cells adjacent to the gray matter and conveys information to the cerebellum about skeletal muscle activity in the limbs generated by UMNs.

SPINOCEREBELLAR SYSTEM LESIONS AND HEREDITARY DISEASES

- *There is a group of **hereditary diseases in which degeneration of spinocerebellar pathways is a prominent feature** (Figure 3–16A). The most common of these is **Friedreich's ataxia**, in which the spinocerebellar tracts, dorsal columns, and corticospinal tracts undergo degeneration.*
- *A sensory ataxia with a positive Romberg sign is a common initial symptom of these diseases, followed by hyporeflexia and altered vibratory sensations.*

SPINAL CORD HEMISECTION AND BROWN-SÉQUARD SYNDROME (SEE FIGURES 3–14, 3–15, AND 3–16B)

- ***Brown-Séquard syndrome*** *results from a lesion or compression of one-half of the spinal cord and may be caused by trauma, a neoplasm, or a herniated disk.*
- *A complete hemisection of the cord results in a lesion of the axons of UMN systems, descending hypothalamic axons, axons in 1 or both dorsal columns, and axons in the spinothalamic and dorsal spinocerebellar tracts.*
- *The hallmark of a lesion to these elements of cord neural systems is that the patient will have 2 ipsilateral long tract signs and 1 contralateral long tract sign.*
- *A lesion of the axons of UMNs, including the corticospinal tract, results in a spastic paresis ipsilateral to and below the level of the injury. A lesion to the fasciculus gracilis or cuneatus results in a loss of joint position sense, tactile discrimination, and vibratory and pressure sensations ipsilateral to and below the lesion.*
- *A lesion of a spinothalamic tract results in a loss of pain and temperature that is contralateral and typically begins 1 or 2 segments below the level of the lesion.*
- *If the hemisection is at cervical spinal cord levels, the patient may have an ipsilateral Horner's syndrome, and at any level above S2 the patient may have a spastic bladder.*

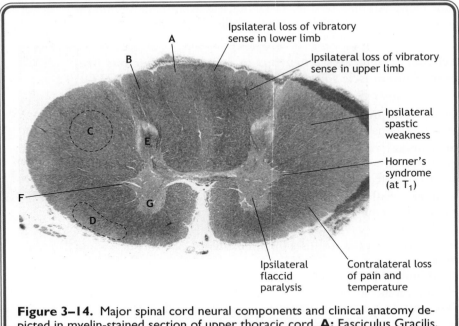

Figure 3–14. Major spinal cord neural components and clinical anatomy depicted in myelin-stained section of upper thoracic cord. **A:** Fasciculus Gracilis. **B:** Fasciculus cuneatus. **C:** Corticospinal tract. **D:** Anterolateral system (spinothalamic tract). **E:** Dorsal horn. **F:** Lateral horn (preganglionic sympathetic neurons). **G:** Ventral horn (lower motor neurons)

- At the level of the lesion, incoming dorsal roots and outgoing ventral roots may be lesioned. Patients may have an ipsilateral loss of all sensation, including touch modalities as well as pain and temperature, and an ipsilateral flaccid paralysis in muscles supplied by the affected spinal segments.

SPINAL SHOCK

- A patient with a **spinal cord hemisection or complete transection** may undergo an initial period of "spinal shock," characterized by flaccid paralysis of muscles innervated by neurons below the transection, loss of all reflexes, and loss of all sensation below the level of the lesion. After days or even weeks, the flaccid paralysis below the lesion is replaced by a spastic weakness and other signs associated with a lesion of axons of UMNs.
- Patients with spinal shock may also have a transient atonic or flaccid bladder, which may become a spastic bladder. In a patient with an atonic bladder, the bladder fills to capacity but urine dribbles through the urethra continuously because the detrusor fails to contract and empty the bladder, and the voluntary urethral sphincter may be weak.
- In patients with a spastic bladder, the bladder contracts in response to a minimum amount of stretch, causing frequent involuntary reflex emptying.

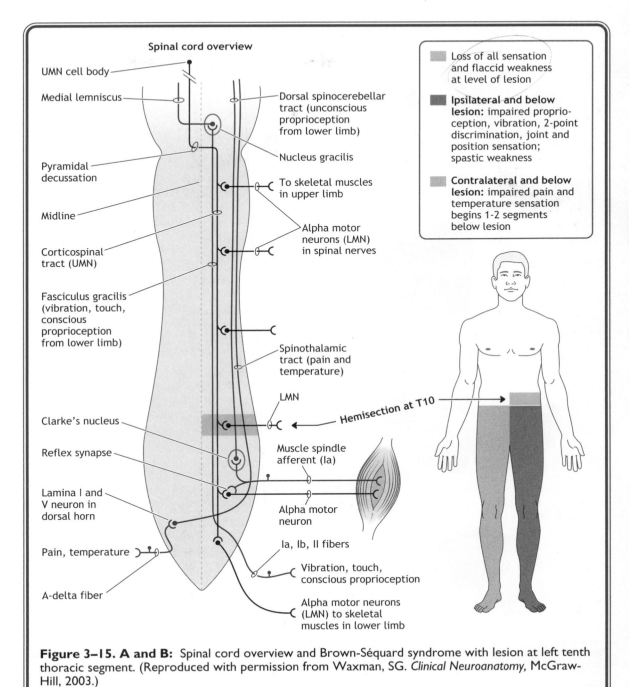

Figure 3–15. A and B: Spinal cord overview and Brown-Séquard syndrome with lesion at left tenth thoracic segment. (Reproduced with permission from Waxman, SG. *Clinical Neuroanatomy*, McGraw-Hill, 2003.)

A Friedreich's ataxia

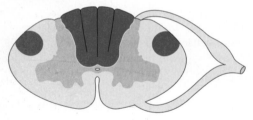

B Brown-Séquard syndrome

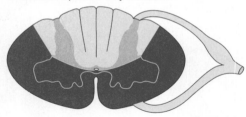

C Anterior spinal artery occlusion

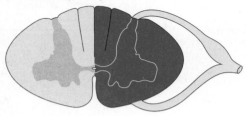

D Subacute combined degeneration

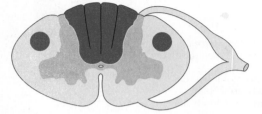

Figure 3–16. A–D: Structures affected in common spinal cord lesions II.

DEFICITS IN PATIENTS AFTER A STROKE INVOLVING THE ANTERIOR SPINAL ARTERY

*In a **sudden occlusion of the anterior spinal artery** (most commonly at thoracic spinal cord levels), the blood supply to the ventrolateral parts of the cord, including the corticospinal tracts and spinothalamic tracts, is disrupted (see Figure 3–16C). Below the level of the lesion (after a period of spinal shock), the patient may have a bilateral spastic paresis and a bilateral loss of pain and temperature.*

SUBACUTE COMBINED DEGENERATION AND THE SPINAL CORD

- ***Subacute combined degeneration** (see Figure 3–16D) is commonly seen in patients with a vitamin B$_{12}$ (cobalamin) deficiency; blood work may reveal antibodies raised against intrinsic factor.*
- *The disease is commonly characterized by swelling of the myelin sheaths in the dorsal columns and corticospinal tracts in the thoracic or cervical cord regions. Patients may have a bilateral spastic paresis and a bilateral alteration of touch, vibration, and pressure sensations below the lesion sites. Peripheral myelin may also be involved, producing paresthesias and impaired sensation in all 4 limbs.*

CLINICAL PROBLEMS

1. Cutting a dorsal root of a spinal nerve may result in:

    A. Hypotonia in skeletal muscles innervated by the cut root

    B. Fasciculations in skeletal muscles innervated by the cut root

    C. Reversed cutaneous reflexes in skin over the denervated muscle

    D. Anterograde degeneration of axons in the spinothalamic tract

    E. A loss of all touch, vibration, and pressure sensations ipsilateral to and below the lesion

2. A neurological exam of your patient reveals a loss of vibratory sense in the lower limb on the right, weakness and hyperactive reflexes in the lower limb on the right, and a loss of pain and temperature that begins below the T8 dermatome on the left. Where is the lesion?

    A. T6 spinal cord segment on the right

    B. T8 spinal cord segment on the right

    C. T10 spinal cord segment on the right

    D. T8 spinal cord segment on the left

    E. T6 spinal cord segment on the left

3. Your patient complains that he burned his hand on his portable heater but did not feel the stimulus. The patient also notes that he has difficulty using either hand. You note that the patient has no response to pinprick in skin of either hand, arm, or

shoulder and a bilateral wasting of intrinsic hand muscles. You suspect that the patient has:

A. Syringomyelia

B. Multiple sclerosis

C. Poliomyelitis

D. Brown-Séquard syndrome

E. Amyotrophic lateral sclerosis

4. A tumor pressing against the dorsal funiculus of the spinal cord at lumbar levels has caused neurological deficits. Your patient might have:

A. A bilateral loss of pain and temperature sensations from sacral dermatomes

B. Altered vibratory sense in the lower limbs

C. A loss of protopathic sensations from thoracic dermatomes

D. Bilateral spastic weakness in the lower limbs

E. A loss of reflexes in the upper limbs

5. The spinal cord has been hemisected at spinal cord segment C5. You might expect retrograde chromatolysis to occur:

A. In dorsal root ganglia contralateral and below the lesion

B. In lamina I of the dorsal horn contralateral to and below the lesion

C. In lamina V of the dorsal horn ipsilateral to and below the lesion

D. In the precentral gyrus ipsilateral to the lesion

E. In Clarke's nucleus contralateral to and below the lesion

6. Your patient has received a painful wound.

All of the following neural structures are active in the suppression of pain except:

A. Neurons in the periaqueductal gray of the midbrain

B. Neurons in lamina II

C. Neurons in the brainstem

D. Medial division dorsal root fibers

E. Neurons in lamina V

7. A weight lifter is attempting to set a new personal best in competition. As the lifter attempts to raise the weight to his chest by flexing his forearms at the elbow, he suddenly drops the weight to the floor. Which neurons were excessively stimulated and caused the lifter to drop the weight?

A. Renshaw cells

B. Lateral reticulospinal tract neurons

    C. Gamma motor neurons

    D. Class Ib afferent neurons

    E. Class II afferent neurons

8. You are evaluating your adult patient's muscle strength and muscle stretch reflexes. Which of the following would be expected in a normal motor exam?

    A. A positive Romberg sign

    B. Absent abdominal reflexes

    C. A grade of 2+ for reflexes

    D. A grade of muscle strength of 3/5

    E. Clonus

9. A construction worker falls off a ladder and fractures a vertebra. A neurological exam conducted 2 weeks after the accident reveals that the individual has a complete hemisection of the right side of the spinal cord at T5. In this patient, you might expect a pain and temperature loss:

    A. In the upper and lower limbs on the left

    B. In all dermatomes below T7 on the left

    C. In the T3 dermatome on the right

    D. Below the T5 dermatome on the right

    E. That is caused by degeneration of dorsal root ganglion cells below T7 on the left.

10. You would expect a spastic weakness in this patient to be:

    A. Ipsilateral to and below the level of the lesion

    B. Ipsilateral to and above the level of the lesion

    C. Observed before the onset of spinal shock

    D. Contralateral to and below the level of the lesion

    E. Bilateral to and at the level of the lesion

11. A newborn infant has difficulty sucking, swallowing, or breathing and has flaccid weakness in the limbs. What kind of motor disorder might the patient have?

    A. Poliomyelitis

    B. Amyotrophic lateral sclerosis

    C. Werdnig-Hoffmann disease

    D. Guillain-Barré syndrome

    E. Myasthenia gravis

12. The anterior spinal artery has been occluded at the point where it supplies the midthoracic segments of the spinal cord. What might you expect your patient to have?

    A. Spastic weakness of both upper limbs

    B. Altered touch sensations in both lower limbs

    C. Hyporeflexia in both lower limbs

    D. Bilateral degeneration of LMNs in the ventral horns of the lumbar spinal cord

    E. Bilateral Babinski signs

13. An anesthetic has been administered to a patient. Which fibers will be anesthetized first?

    A. Class Ib fibers

    B. A-delta fibers

    C. Gamma motor neurons

    D. Preganglionic autonomic axons

    E. C fibers

14. Your patient has suffered trauma to the spinal cord. During a period of spinal shock, what might you expect to observe in the patient?

    A. Hyperactive reflexes below the lesion

    B. Flaccid weakness below the lesion

    C. A spastic bladder

    D. A clasp knife reflex

    E. A Babinski sign

15. A tumor is beginning to compress the right side of the spinal cord at the C3 spinal cord segment and is impinging on the corticospinal and spinothalamic tracts. What might you expect to observe first in the patient?

    A. Alteration of pain and temperature sensations from the left upper limb

    B. Alteration of pain and temperature sensations from the left lower limb

    C. Spastic weakness of the left upper limb

    D. Spastic weakness of the left lower limb

    E. Bilateral loss of pain and temperature sensations from sacral dermatomes

16. Which of the following diseases or lesions is <u>incorrectly</u> matched with a location where neuronal cell bodies might show degeneration or retrograde changes as a result of the disease process?

    A. Tabes dorsalis/neurons in dorsal root ganglia

    B. Poliomyelitis/neurons in lamina IX

C. Syringomyelia/neurons in lamina II

D. Brown-Séquard syndrome/neurons in dorsal root ganglia

E. Lesion of the dorsal spinocerebellar tract/Clarke's nucleus

17. A Navy officer who recently retired from active duty overseas develops sharp pains in both legs, altered vibratory and pressure sensations in both legs, and an increased tendency to urinate. When he walks out to get the morning newspaper, his gait is unsteady. Strength is 5/5 in both lower limbs, but the Achilles tendon reflexes are graded as 1 bilaterally. Plantar stimulation yields a flexor response. An evaluation of cranial nerves reveals that the patient's pupils constrict in response to a near stimulus but not in response to light. What might the patient have?

A. Poliomyelitis

B. Amyotrophic lateral sclerosis

C. Tabes dorsalis

D. Syringomyelia

E. Vitamin B_{12} deficiency

18. What else might your neurological exam of the patient in Question 17 reveal?

A. Wasting of lower limb muscles

B. Patient will sway back and forth with eyes closed

C. Horner's syndrome

D. A central canal syrinx

E. Absence of abdominal and cremasteric reflexes

19. You are a pathologist examining a spinal cord acquired during an autopsy. You section the cord and stain the sections with a myelin stain. Which of the following would you normally expect to observe in a stained section?

A. The least amount of staining in the white matter would be in Lissauer's tract.

B. Dorsal and ventral roots would be unstained.

C. The class C fibers in the spinothalamic tracts would be unstained.

D. The lateral division of the dorsal root would be more heavily stained than the medial division of the dorsal root.

E. The axons in the fasciculus gracilis would be more heavily stained than those in the fasciculus cuneatus.

20. A middle-aged man develops motor weakness in both the upper and lower limbs. The thenar eminences of both hands appear to have atrophied, and the biceps and triceps tendon reflexes are diminished. In the weak lower limbs, the muscle stretch reflexes are brisk. There are no sensory deficits. Which of the following conclusions might be made?

A. The patient may have suffered from an occlusion of the anterior spinal artery at cervical spinal cord levels.

B. The patient has Brown-Séquard syndrome.

C. The patient has Werdnig-Hoffmann syndrome.

D. The patient has Guillain-Barré syndrome.

E. The patient has amyotrophic lateral sclerosis.

MATCHING PROBLEMS

Questions 21–30: Fiber match

Match the following fiber types with their function or descriptive feature.

Choices (each choice may be used once, more than once, or not at all):

A. A-delta fibers

B. A-beta fibers

C. C fibers

D. Ia fibers

E. Gamma motor neurons

F. Alpha motor neurons

G. Preganglionic autonomic axons

H. Postganglionic autonomic axons

I. More than one choice in A–H is correct

21. Fibers in a dorsal root that are most susceptible to anesthesia.

22. Fibers associated with the modalities of burning pain and warmth.

23. Fibers that make muscle spindles more sensitive to stretch.

24. Sensory fibers most resistant to anesthesia.

25. Fibers that synapse in Rexed lamina II.

26. Fibers derived from neural crest cells.

27. Neuron cell bodies of these fibers are found in Rexed lamina IX.

28. Innervate sharp pain receptors.

29. Slowest conducting motor fibers.

30. Sensory fibers most sensitive to reduced oxygen levels.

Questions 31–42: Cell body/ axon terminal match

Choices: (each choice may be used once, more than once, or not at all):

A. Lamina I of spinal cord

B. Lamina II of spinal cord

C. Lamina IX of spinal cord

D. Lamina VII of spinal cord

E. In a ganglion outside the CNS

F. Inside the CNS but outside the spinal cord

Using the choices A–F above, indicate the location

31. of the neuronal cell bodies of dorsal spinocerebellar tract axons.

32. of the axon terminals of Renshaw cells.

33. of the neuronal cell bodies of gamma motor axons.

34. of the cell bodies of nucleus cuneatus axons.

35. of the neuronal cell bodies of afferents that innervate Golgi tendon organs.

36. of the neuronal cell bodies of preganglionic sympathetic axons.

37. of the axon terminals of fasciculus gracilis fibers.

38. of axon terminals of the corticospinal tract.

39. of axon terminals of group Ia fibers in a stretch reflex.

40. of the neuronal cell bodies of anterolateral system fibers.

41. of axon terminals of A-delta fibers.

42. of neuronal cell bodies derived from neural crest cells.

Questions 43–58: For each symptom or symptoms in the following list, indicate the spinal cord disease or lesion with which it is most likely associated.

Choices (each choice may be used once, more than once, or not at all):

A. Syringomyelia

B. Brown–Séquard syndrome

C. Amyotrophic lateral sclerosis

D. Poliomyelitis

E. Tabes dorsalis

F. Complete transection of spinal cord

G. Multiple sclerosis

H. Guillain Barré syndrome

I. Occlusion of the anterior spinal artery at the level of the T5 spinal cord segment.

43. Patient has bilateral hyporeflexia, muscle weakness, and fasciculations; no sensory deficits.

44. Patient has an ipsilateral loss of epicritic sensations and contralateral loss of pain and temperature below the lesion.

45. Patient has a bilateral loss of protopathic sensations and a bilateral flaccid paralysis in the upper limbs; vibratory sense intact.

46. Patient has a bilateral alteration of pressure and vibratory sensations and hyporeflexia; muscle strength is normal and patient is hypersensitive to pain.

47. Patient has bilateral spastic paresis and bilateral loss of pain and temperature sensations in the lower limbs; vibratory sense is intact.

48. Patient has bilateral compression of axons from lamina I and lamina V neurons.

49. Patient has signs of spinal shock initially below a lesion; bilateral Babinski signs are evident later.

50. Patient has inflammation of myelinated axons in white matter.

51. Patient has hyporeflexia in the upper limbs and hyperreflexia in the lower limbs; all four limbs are weak.

52. Patient has pain and paresthesias in the lower limbs and urine retention; no weakness.

53. Patient has inflammation of myelin produced by derivatives of neural crest cells.

54. Patient has a loss of pain and temperature sensations contralateral and below a lesion, a loss of touch and pressure sensations ipsilateral and below a lesion, and a Babinski sign ipsilateral and below the lesion.

55. Patient cannot tell the temperature of water on his hands when he washes them; fasciculations are evident in both of his forearms, and patient cannot hold an object with either hand.

56. Patient has bilateral wasting of the thenar eminences and suppressed reflexes in upper limbs; in both lower limbs, muscle strength is graded at 3 and muscle stretch reflexes graded at 4.

57. Patient has fasciculations and atrophy of skeletal muscles in the legs, and bilateral suppressed Achilles tendon reflexes.

58. Patient sways with his eyes closed and has pupillary light reflex anomalies; he walks with a broad-based gait, but cannot feel his legs.

ANSWERS

1. The answer is A. Hypotonia in skeletal muscles innervated by the cut root results because of the loss of the Ia fibers providing the sensory limb of the muscle stretch reflex. Choices B and C are seen with UMN lesions, spinothalamic axons are second neurons not affected by severing a dorsal root, and choice E would result from a dorsal column lesion.

2. The answer is A. The patient has a hemisection that gives rise to 2 right-sided (ipsilateral) long tract signs and a loss of pain and temperature that begins on the left (contralateral) and 2 segments below the lesion.

3. The answer is A. Patients with multiple sclerosis (choice B) have sensory or motor deficits separated in time and space that are frequently seen with optic neuritis. Choices C and E are pure motor neuron diseases.

4. The answer is B. The tumor has compressed both of the fasciculus gracilis components of the dorsal columns. Compression of the lateral funiculus might cause the signs in choices A and D, and choice E would be evident if the tumor compressed the fasciculus cuneatus at cervical levels.

5. The answer is B. Axons of these neurons cross below the lesion and course in the spinothalamic tract. Choice C would also show chromatolysis contralateral to and below the lesion. Chromatolysis would be seen ipsilaterally in choices A and E and contralaterally in choice D.

6. The answer is E. Choices A, B, and C contain neurons that suppress pain from the brainstem or locally in the dorsal horn. Choice D fibers stimulate spinothalamic neurons to respond to a nonnoxious stimulus.

7. The answer is D. The clasp knife reflex has been activated by a rapid buildup of muscle force, causing a large discharge by the GTOs and Ib afferents and a sudden inhibition of the affected muscles.

8. The answer is C. A positive Romberg sign results from a dorsal column lesion. Absent abdominal reflexes and clonus are seen in patients with UMN lesions. Normal strength should be 5/5, at which the patient generates movement against full resistance.

9. The answer is B. In a spinal cord hemisection, loss of pain and temperature begins 1–2 segments contralateral and below the lesion.

10. The answer is A. A lesion of axons of UMNs in the spinal cord results in a spastic weakness, a long tract sign, that is ipsilateral to and below the lesion.

11. The answer is C. Werdnig-Hoffmann disease, or infantile spinal muscular atrophy, results in destruction of LMNs in infants or young children. Infants may have difficulty sucking, swallowing, or breathing and weakness in the limbs; they are "floppy" babies.

12. The answer is E. The anterior spinal artery supplies the ventrolateral parts of the cord, including the corticospinal tracts and spinothalamic tracts. Below the level of the lesion at midthoracic levels, the patient may have a bilateral spastic paresis and Babinski signs and a bilateral loss of pain and temperature.

13. The answer is E. C fibers are the smallest diameter axons in either the dorsal and ventral root and the most susceptible to anesthesia. All other choices are larger diameter, myelinated fibers.

14. The answer is B. A patient with an initial period of spinal shock, after spinal cord trauma, has flaccid paralysis of muscles innervated by neurons below the lesion, loss of all reflexes, and loss of all sensation below the level of the lesion. After days or even weeks, the flaccid paralysis below the lesion is replaced by a spastic weakness and other signs associated with an lesion of axons of UMNs (choices A, D, and E) and a spastic bladder (choice C).

15. The answer is B. The crossed pain and temperature information from sacral and lumbar dermatomes is conveyed by laterally placed spinothalamic axons at cervical spinal cord levels and is the first of the choices given to be compressed by the tumor.

16. The answer is C. The crossing spinothalamic axons compressed by the syrinx arise from laminae I and V. Lamina II neurons receive C fiber input and act to suppress pain.

17. The answer is C. Tabetic patients commonly have the 3 "Ps," paresthesias, pain, and polyuria, which result from a bilateral degeneration of the dorsal columns. Paresthesias result from impaired vibration and position sense in the lower limbs. The radiating pain results from hypersensitivity of the small-diameter A-delta and class C pain and temperature dorsal roots. Polyuria or urine retention may also be evident. Patients with tabes dorsalis may also have Argyll Robertson pupils, which accommodate, but are unreactive to, light.

18. The answer is B. A lesion of the dorsal columns may result in a sensory ataxia and a positive Romberg sign.

19. The answer is A. The least amount of staining in the white matter would be in Lissauer's tract because this tract contains thinly myelinated A-delta and class C fibers. Both dorsal and ventral roots would be stained, the medial division would stain more heavily than the lateral division, the dorsal columns would stain equally, and there are no class C fibers in the spinothalamic tracts.

20. The answer is E. ALS is a pure motor disease that has a combination of UMN and LMN signs and no sensory deficits. Choices A and B have both motor and sensory signs, and choices C and D affect only LMNs.

21. C	34. E	47. I
22. C	35. E	48. A
23. E	36. D	49. F
24. D	37. F	50. G
25. C	38. C	51. C
26. I (A–D, H apply)	39. C	52. E
27. I (E and F apply)	40. A	53. H
28. A	41. A	54. B
29. H	42. E	55. A
30. D	43. D	56. C
31. D	44. B	57. D
32. C	45. A	58. E
33. C	46. E	

CHAPTER 4
BRAINSTEM

I. **The brainstem consists of the midbrain, the pons, and the medulla; 9 of the 12 cranial nerves enter or exit from parts of the brainstem (Figure 4–1A, B, Table 4–1).**

    A. The **midbrain is the most rostral part of the brainstem**; the oculomotor and trochlear nerves (cranial nerves [CNs] III and IV) arise from the midbrain (see Figures 4–2A, 4–2B, 4–3A, and 4–3B).

        1. On its dorsal surface, the midbrain contains **a pair of superior colliculi and a pair of inferior colliculi**.

            a. The **superior colliculi** function in conjugate gaze.

            b. The **inferior colliculi** are central auditory structures.

            c. **Axons of the trochlear nerve** emerge from the midbrain just inferior to the inferior colliculus after crossing in the superior medullary velum.

        2. On its ventral surface, the midbrain contains **2 cerebral peduncles separated by an interpeduncular fossa (Figures 4–2A and 4–2B)**.

            a. Each peduncle contains **corticopontine, corticospinal, and corticobulbar axons**, which synapse with pontine nuclei, lower motor neurons of spinal nerves, and lower motor neurons of cranial nerves, respectively.

            b. Each peduncle contains a **substantia nigra** that is a source of central nervous system (CNS) neurons that use dopamine and gamma-aminobutyric acid (GABA) as neurotransmitters.

            c. Fibers of the **oculomotor nerve** emerge from the midbrain between the peduncles in the interpeduncular fossa.

        3. The midbrain contains the **cerebral aqueduct**, which interconnects the third and fourth ventricles; the periaqueductal gray matter surrounds the cerebral aqueduct.

THE CEREBRAL AQUEDUCT AND THE PERIAQUEDUCTAL GRAY

- The **cerebral aqueduct** is a common site of an obstruction that results in a **noncommunicating hydrocephalus**.
- The **periaqueductal gray** contains neurons with opioid receptors that are stimulated by the spinomesencephalic component of the anterolateral system; these neurons function in the suppression of pain.

    B. The **pons** is between the midbrain and the medulla; the trigeminal, abducens, facial, and vestibulocochlear nerves (CNs V, VI, VII, and VIII) enter or exit from ventral aspect of the pons (see Figures 4–1A, 4–4A, 4–4B, 4–5A, and 4–5B).

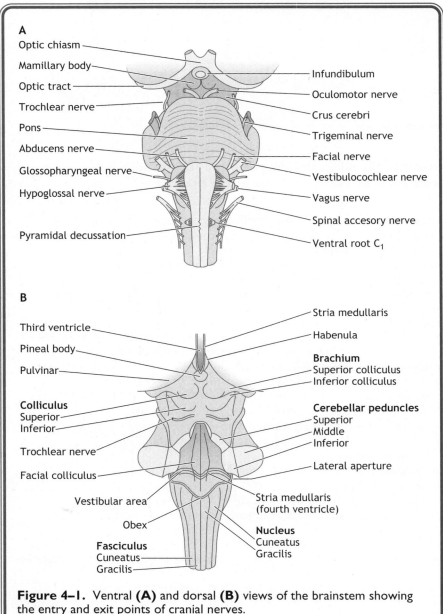

A

Optic chiasm
Mamillary body
Optic tract
Trochlear nerve
Pons
Abducens nerve
Glossopharyngeal nerve
Hypoglossal nerve

Pyramidal decussation

Infundibulum
Oculomotor nerve
Crus cerebri
Trigeminal nerve
Facial nerve
Vestibulocochlear nerve
Vagus nerve
Spinal accesory nerve
Ventral root C_1

B

Third ventricle
Pineal body
Pulvinar

Colliculus
Superior
Inferior

Trochlear nerve
Facial colliculus
Vestibular area
Obex

Fasciculus
Cuneatus
Gracilis

Stria medullaris
Habenula
Brachium
Superior colliculus
Inferior colliculus

Cerebellar peduncles
Superior
Middle
Inferior

Lateral aperture

Stria medullaris
(fourth ventricle)

Nucleus
Cuneatus
Gracilis

Figure 4–1. Ventral **(A)** and dorsal **(B)** views of the brainstem showing the entry and exit points of cranial nerves.

Table 4–1. Cranial nerves: functional and clinical features.

No.	Name	Type	Function	Lesions Result In
I	Olfactory	Sensory	Smells	Anosmia
II	Optic	Sensory	Sees	Visual field deficits (anopsia) Loss of light reflex with CN III
VIII	Cochlear vestibular	Sensory	Hears Linear acceleration (gravity) Angular acceleration (head turning)	Sensorineural hearing loss Loss of balance Nystagmus (fast phase away from lesion)
III	Oculomotor	Motor	To all eyeball muscles except LR and SO Adduction (medial rectus) most important action Elevates upper eyelid (levator palpebrae superioris) Constricts pupil (sphincter pupillae) Accommodates (ciliary muscle)	Diplopia, external strabismus Loss of parallel gaze Ptosis Dilated pupil, loss of motor limb of light reflex with CN II Loss of near response
IV	Trochlear	Motor	To superior oblique; depresses and abducts eyeball Intorts	Weakness looking down when eye adducted Difficulty reading, going down stairs Head tilts away from lesioned side
VI	Abducens	Motor	To lateral rectus; abducts eyeball	Diplopia, internal strabismus Loss of parallel gaze, "pseudo-ptosis"
XI	Accessory	Motor	Turns head to opposite side (sternocleidomastoid) Elevates and rotates scapula (trapezius)	Weakness turning head to opposite side Shoulder droop, difficulty combing hair
XII	Hypoglossal	Motor	To all muscles that act on tongue except palatoglossus (X) (hyoglossus, styloglossus, genioglossus, intrinsics)	Tongue deviation on protrusion toward lesioned nerve Dysarthria
V	Trigeminal ophthalmic (V1)	Mixed	General sensation (touch, pain, temperature) of forehead/scalp/cornea	Loss of general sensation in skin of forehead/scalp Loss of sensory limb of blink reflex with CN VII

(continued)

Table 4–1. Cranial nerves: functional and clinical features. (*Continued*)

No.	Name	Type	Function	Lesions Result In
	Maxillary (V2)		General sensation of palate, nasal cavity, maxillary face, maxillary teeth	Loss of general sensation in skin over maxilla, maxillary teeth
	Mandibular (V3)		General sensation of anterior 2/3 tongue, mandibular face, mandibular teeth Motor to muscles of mastication (temporalis, masseter, medial and lateral pterygoids) plus anterior belly of digastric, mylohyoid, tensor tympani, tensor palati	V3, loss of general sensation in skin over mandible, mandibular teeth, tongue, weakness in chewing Jaw deviation toward lesioned nerve
VII	Facial	Mixed	To muscles of fascial expression, including orbicularis oculi and oris, platysma, buccinator, stapedius Tastes anterior 2/3 tongue, soft palate Salivates (submandibular, sublingual glands) Tears (lacrimal gland) Makes mucus (nasal and palatine glands)	Corner of mouth droops, unable to close eye, wrinkle forehead Loss of motor limb of blink reflex Hyperacusis Alteration or loss of taste Reduction in output of saliva Eye dry Reduction in secretions
IX	Glosso-pharyngeal	Mixed	General sensation of oropharynx, carotid sinus, and carotid body Taste and general sensation posterior 1/3 of tongue Motor to one muscle: stylo-pharyngeus Salivates (parotid gland)	Loss of sensory limb of gag reflex with CN X Reduction in output of saliva
X	Vagus	Mixed	To muscles of palate except tensor palati (V) To all muscles of pharynx except stylopharyngeus (X) To all muscles of larynx Senses larynx and laryngo-pharynx To glands and smooth muscle in thorax, foregut, and midgut	Palate droop, uvula deviation away from lesioned nerve Dysphagia, loss of motor limb of gag reflex with CN IX Hoarseness, dysphonia
	Descending hypothala-mic axons	Motor	Elevates upper eyelid (superior tarsal muscle) Dilates pupil Innervates sweat glands of face and scalp	Ptosis Constricted pupil (miosis), Loss of sweating (anhidrosis) (Horner's syndrome)

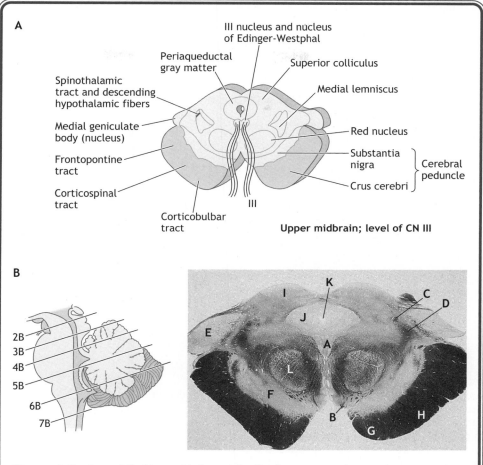

A

III nucleus and nucleus
of Edinger-Westphal

Periaqueductal
gray matter

Superior colliculus

Spinothalamic
tract and descending
hypothalamic fibers

Medial lemniscus

Medial geniculate
body (nucleus)

Red nucleus

Frontopontine
tract

Substantia
nigra

Crus cerebri

} Cerebral
peduncle

Corticospinal
tract

Corticobulbar
tract

III

Upper midbrain; level of CN III

B

2B
3B
4B
5B
6B
7B

Figure 4–2. **A and B:** Upper Midbrain. **A:** Oculomotor nucleus and nucleus of Edinger-Westphal. **B:** Oculomotor nerve fibers. **C:** Spinothalamic tract/descending hypothalamic fibers. **D:** Medial lemniscus. **E:** Medial geniculate nucleus. **F:** Substantia nigra. **G:** Corticobulbar tract. **H:** Corticospinal tract. **I:** Superior colliculus. **J:** Periaqueductal gray matter. **K:** Cerebral aqueduct. **L:** Red nucleus. Inset: Levels of sections in Figures 4–2 to 4–7. (**A:** Reproduced with permission from Waxman, SG. *Clinical Neuroanatomy.* McGraw-Hill, 2003.)

1. The **trigeminal nerve fibers** enter or exit through the rostral pons; **fibers of the abducens, facial, and vestibulocochlear nerves** enter or exit at the pontomedullary junction.
2. Ventrally, the **pons is enlarged by pontine nuclei and their axons**, which enter the cerebellum in the middle cerebellar peduncle.
3. Dorsally, the **pons is overlain by the cerebellum** and is separated from it by the fourth ventricle.

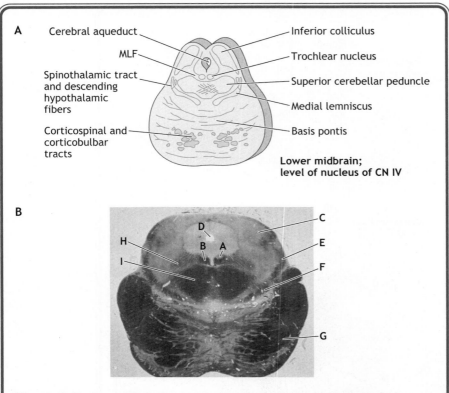

A

Cerebral aqueduct

MLF

Spinothalamic tract
and descending
hypothalamic
fibers

Corticospinal and
corticobulbar
tracts

Inferior colliculus

Trochlear nucleus

Superior cerebellar peduncle

Medial lemniscus

Basis pontis

Lower midbrain;
level of nucleus of CN IV

B

Figure 4–3. A and B: Lower midbrain. **A:** Trochlear nucleus. **B:** Medial longitudinal fasciculus. **C:** Inferior colliculus. **D:** Cerebral aqueduct. **E:** Spinothalamic tract and descending hypothalamic fibers. **F:** Medial lemniscus. **G:** Corticospinal/corticobulbar tracts. **H:** Central tegmental tract. **I:** Crossed axons of superior cerebellar peduncle. (**A:** Reproduced with permission from Waxman, SG. *Clinical Neuroanatomy,* McGraw-Hill, 2003.)

C. The **medulla** is caudal to the pons and is continuous with the spinal cord; fibers of the glossopharyngeal, vagus, and hypoglossal nerves (CNs IX, X, and XII) enter or exit ventrally from the medulla (see Figures 4–1A, 4–1B, 4–6A, 4–6B, 4–7A, and 4–7B).

　1. Near the ventral midline of the caudal medulla are the **medullary pyramids,** which contain the **corticospinal and corticobulbar tracts.**

　2. Dorsolateral to each pyramid is an **olive,** which **forms a pair of swellings in the rostral medulla.**

　3. The **roots of the glossopharyngeal and vagus nerves** enter or exit just dorsolateral to the olive; the **roots of hypoglossal nerve** exit medial to the olive and lateral to each pyramid.

　4. The **inferior part of the fourth ventricle** is dorsal to the rostral part of the medulla; the **caudal medulla contains the central canal.**

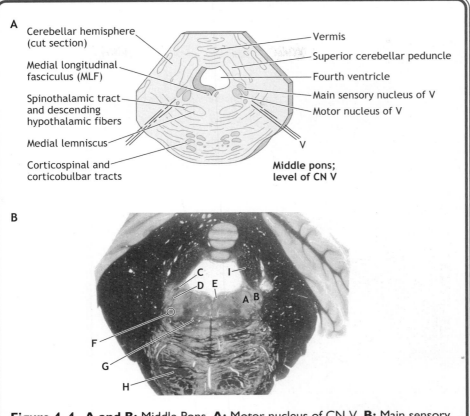

Figure 4–4. A and B: Middle Pons. **A:** Motor nucleus of CN V. **B:** Main sensory nucleus of CN V. **C:** Mesencephalic tract of CN V. **D:** Fibers of CN V. **E:** Medial longitudinal fasciculus. **F:** Spinothalamic tract and descending hypothalamic fibers. **G:** Medial lemniscus. **H:** Corticospinal tract. **I:** Superior cerebellar peduncle. (**A:** Reproduced with permission from Waxman, SG. *Clinical Neuroanatomy.* McGraw-Hill, 2003.)

     5. The **caudal medulla also contains the rostral parts of the dorsal columns and the dorsal column nuclei** (nucleus cuneatus and gracilis).

II. The midbrain, pons, and medulla contain components of motor and sensory systems and fiber bundles associated with them (Figures 4–2 to 4–7).

    A. Corticospinal and corticobulbar tracts

       1. The **corticospinal tracts** that contain axons of upper motor neurons course ventrally and medially through the length of the midbrain, pons, and medulla.

       2. The corticospinal tracts decussate near the spino-medullary junction and descend into the spinal cord to innervate lower motor neurons in spinal nerves (Figure 4–7B).

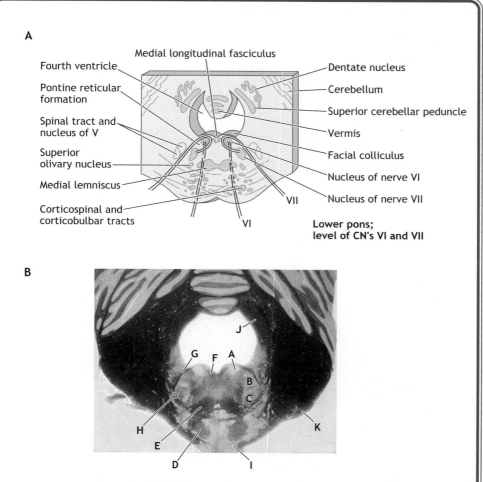

A

Medial longitudinal fasciculus

Fourth ventricle

Pontine reticular formation

Spinal tract and nucleus of V

Superior olivary nucleus

Medial lemniscus

Corticospinal and corticobulbar tracts

Dentate nucleus

Cerebellum

Superior cerebellar peduncle

Vermis

Facial colliculus

Nucleus of nerve VI

Nucleus of nerve VII

VII

VI

Lower pons; level of CN's VI and VII

B

Figure 4–5. A and B: Lower Pons. **A:** Abducens nucleus. **B:** Motor nucleus of CN VII. **C:** Superior olivary nuclei. **D:** Corticospinal/corticobulbar tracts. **E:** Medial lemniscus. **F.** Medial longitudinal fasciculus. **G:** Facial nerve. **H:** Spinothalamic tract and descending hypothalamic fibers. **I:** Pontine nuclei. **J:** Superior cerebellar peduncle. **K:** Middle cerebellar peduncle. (**A:** Reproduced with permission from Waxman, SG. *Clinical Neuroanatomy.* McGraw-Hill, 2003.)

LESION OF CORTICOSPINAL TRACT AXONS

A *lesion of corticospinal tract axons in the brainstem above the decussation* may result in a *spastic hemiparesis* in the limbs contralateral and below the lesion.

   3. **Corticobulbar axons course with corticospinal axons and innervate lower motor neurons** in cranial nerve nuclei (see later discussion).

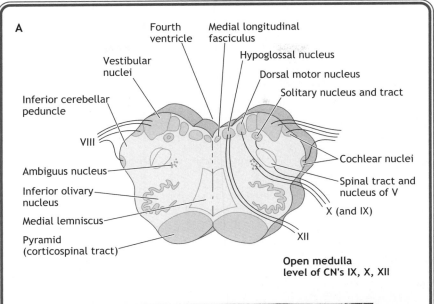

A

Fourth
ventricle

Medial longitudinal
fasciculus

Hypoglossal nucleus

Dorsal motor nucleus

Vestibular
nuclei

Solitary nucleus and tract

Inferior cerebellar
peduncle

VIII

Cochlear nuclei

Ambiguus nucleus

Spinal tract and
nucleus of V

Inferior olivary
nucleus

X (and IX)

Medial lemniscus

XII

Pyramid
(corticospinal tract)

**Open medulla
level of CN's IX, X, XII**

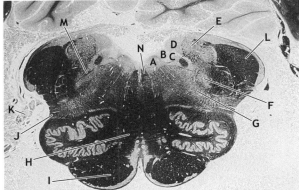

B

Figure 4–6. A and B: Open Medulla. **A:** Hypoglossal nucleus. **B:** Dorsal motor nucleus of X. **C:** Solitary nucleus and tract (black dot). **D:** Medial vestibular nucleus. **E:** Inferior vestibular nucleus. **F:** Spinal nucleus of V. **G:** Nucleus ambiguus. **H:** Medial lemniscus. **I:** Corticospinal tract. **J:** Spinothalamic tract and descending hypothalamic fibers. **K:** Roots of CN IX or CN X. **L:** Inferior cerebellar peduncle. **M:** Fibers of CNs VII, IX, or X forming solitary tract. **N.** Medial longitudinal fasciculus. (**A:** Reproduced with permission from Waxman, SG. *Clinical Neuroanatomy.* McGraw-Hill, 2003.)

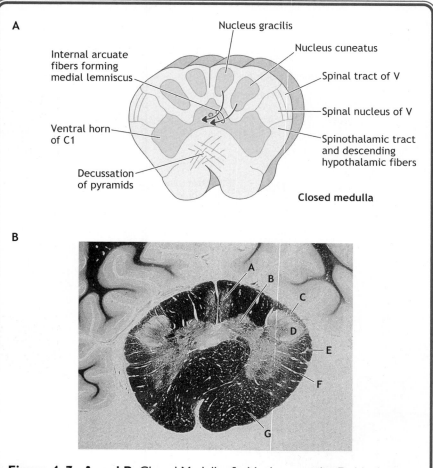

Figure 4–7. A and B: Closed Medulla. **A:** Nucleus gracilis. **B:** Nucleus cuneatus. **C:** Spinal tract of V. **D:** Spinal nucleus of V. **E:** Dorsal spinocerebellar tract. **F:** Spinothalamic tract and descending hypothalamic fibers. **G:** Corticospinal tract and decussation. (**A:** Reproduced with permission from Waxman, SG. *Clinical Neuroanatomy.* McGraw-Hill, 2003.)

B. Components of the dorsal column/medial lemniscal system
   1. The **dorsal column/medial lemniscal system conveys the modalities of touch, vibration, and pressure from the limbs, neck, and trunk**.
   2. The dorsal columns (fasciculus gracilis and cuneatus) **end by synapsing in the dorsal column nuclei** (nucleus gracilis and cuneatus) in the dorsal part of the caudal medulla (Figures 4–1B, 4–7A, and 4–B).
   3. **Axons from the 2 dorsal column nuclei cross immediately in the vicinity of the nuclei as internal arcuate fibers and form the medial lemniscus** on the contralateral side of the medulla.

4. The **medial lemniscus courses medially through the medulla dorsal** to the corticospinal tract and then **moves dorsally and laterally** as it courses through the pons and midbrain.
5. The medial lemniscus **projects to the ventral posterior lateral nucleus (VPL) of the thalamus**.

COMPLETE LESION OF THE MEDIAL LEMNISCUS

*A **complete lesion of the medial lemniscus in the brainstem** may result in a **loss of touch, vibration, and pressure sensations** in the upper and lower limb, neck, and trunk contralateral and below the lesion.*

    C. **The anterolateral system**
       1. The **spinothalamic tract and other components of the anterolateral system convey pain and temperature sensations** from the limbs, neck, and trunk.
       2. The spinothalamic tract has its **cells of origin in the dorsal horn of the contralateral spinal cord** and courses laterally through the medulla, pons, and midbrain.
       3. The spinothalamic tract **joins with the medial lemniscus in the rostral pons and projects to the VPL nucleus of the thalamus (Figures 4–2A, 4–2B, 4–3A, and 4–3B)**.

LESION OF THE SPINOTHALAMIC TRACT

*A **lesion of the spinothalamic tract in the brainstem** may result in a **loss of pain and temperature sensations** from the upper and lower limb, neck, and trunk contralateral and below the lesion.*

    D. **Descending hypothalamic fibers**
       1. Descending hypothalamic fibers course through the lateral part of the brainstem with the spinothalamic tract (Figures 4–5A, 4–5B, 4–6A, and 4–6B).
       2. The descending hypothalamic fibers arise from the hypothalamus, course without crossing in the lateral part of the brainstem, and descend into the spinal cord.
       3. The descending hypothalamic fibers **innervate preganglionic sympathetic and parasympathetic neurons** in the brainstem and spinal cord.

HORNER'S SYNDROME

- *Patients with a **lesion of the descending hypothalamic axons** in the brainstem may have a central **Horner's syndrome** that is always ipsilateral to the side of the lesion. Patients with Horner's syndrome have **miosis, ptosis, and anhidrosis**.*
- *Patients with a central Horner's syndrome may also have a **loss of pain and temperature sensations** in the limbs and trunk contralateral to the lesion due to of the proximity of the spinothalamic tract to the descending hypothalamic axons.*

    E. **The medial longitudinal fasciculus (MLF) (see Figures 4–2 to 4–7)**
       1. The medial longitudinal fasciculus courses through the length of the brainstem in the floor of the fourth ventricle, adjacent to the central canal in the medulla and adjacent to the cerebral aqueduct in the midbrain.
       2. The MLF links the vestibular nuclei and centers for conjugate gaze with the abducens, trochlear, and oculomotor nuclei.

THE MLF, MULTIPLE SCLEROSIS, AND NEUROSYPHILIS

- *The MLF is particularly susceptible to CNS diseases such as **multiple sclerosis and neurosyphilis**.*
- *Lesions of the medial longitudinal fasciculus may result in **internuclear ophthalmoplegia**, which disrupts horizontal conjugate gaze and the vestibulo-ocular reflex (see Chapter 7).*

    F. Cerebellar peduncles (Figures 4–1B and 4–5B)
       1. The **superior cerebellar peduncle** is in the midbrain and mainly conveys axons out of the cerebellum from the deep cerebellar nuclei.
       2. The **middle cerebellar peduncle** is in the pons and conveys axons from the pontine nuclei into the cerebellum.
       3. The **inferior cerebellar peduncle** conveys axons into and out of the cerebellum from the spinal cord and brainstem.

III. The reticular formation forms the core of the brainstem.

    A. The reticular formation **contains monoaminergic and cholinergic neurons** that project to wide areas of the CNS.

    B. Monoaminergic and cholinergic neurons form the **ascending arousal system** that projects to the diencephalon and cortex, regulate levels of arousal and consciousness, and coordinate nonrapid eye movement (non-REM) and REM sleep (see Chapter 8).

    C. The reticular formation **receives pain and temperature information** from the anterolateral system and projects to the intralaminar nuclei of the thalamus.

    D. The **paramedian pontine reticular formation** (PPRF) helps generate horizontal conjugate gaze.

IV. The motor nuclei of cranial nerves are organized into 3 discontinuous longitudinal columns in the medial part of the brainstem (see Figures 4–8 and 4–9).

    A. The **most medial column of motor nuclei** contains lower motor neurons of cranial nerves III, IV, VI, and XII that innervate ocular muscles and tongue muscles.
       1. The **oculomotor nuclei** are located in the midbrain at the level of the superior colliculus.
          a. The oculomotor nerve **innervates all of the ocular muscles except the superior oblique and the lateral rectus**.
          b. The oculomotor nerve **innervates the levator palpebrae superioris, which raises the eyelid**.

CN III LESIONS AND EXTERNAL STRABISMUS

- *All of the muscles that adduct the eyeball are innervated by the oculomotor nerve, so that a **lesion to the oculomotor nucleus or nerve** may result in a **laterally deviated eyeball (an external strabismus)** because of an inability to adduct the eyeball.*
- *Patients may have a **ptosis** resulting from weakness of the levator palpebrae superioris.*
- *Lesions of CN III may also result in **parasympathetic deficits** (see later discussion of Edinger-Westphal nucleus).*

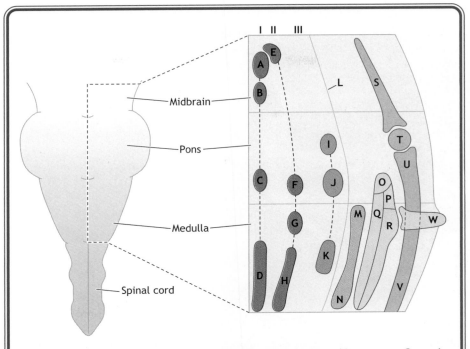

Figure 4–8. Longitudinal Overview of the Organization of Brainstem Cranial Nerve Nuclei. I, II, and III represent the discontinuous longitudinal columns of motor nuclei. **A:** Oculomotor nucleus. **B:** Trochlear nucleus. **C:** Abducens nucleus. **D:** Hypoglossal nucleus. **E:** Nucleus of Edinger-Westphal. **F:** Superior salivatory nucleus. **G:** Inferior salivatory nucleus. **H:** Dorsal motor nucleus of X. **I:** Motor nucleus of V. **J:** Facial motor nucleus. **K:** Nucleus ambiguus. **L:** Sulcus limitans. **M:** Rostral solitary nucleus. **N:** Caudal solitary nucleus. **O:** Superior vestibular nucleus. **P:** Lateral vestibular nucleus. **Q:** Medial vestibular nucleus. **R:** Inferior vestibular nucleus. **S:** Mesencephalic nucleus of V. **T:** Main, chief, or principal nucleus of V. **U:** Spinal nucleus of V (oral). **V:** Spinal nucleus of V (caudal). **W:** Cochlear nuclei.

2. The **trochlear nuclei** are located in the midbrain between the superior and inferior colliculi (Figures 4–3A and 4–3B).
   a. Axons of the trochlear nerves course dorsomedially, cross in the inferior medullary velum, and exit just caudal to the inferior colliculi.
   b. The trochlear nerve **innervates the superior oblique, which acts to depress, abduct, and intort the eye (Figure 4–9).**

TROCHLEAR NERVE LESIONS AND HEAD TILT

- *Lesions of CN IV result in diplopia when the patient attempts to depress the adducted eye and weakness in the ability to intort the eye.*
- *Patients may tilt their head away from the side of the lesioned CN IV to reduce the diplopia that results from an unopposed extorsion by muscles innervated by CN III.*

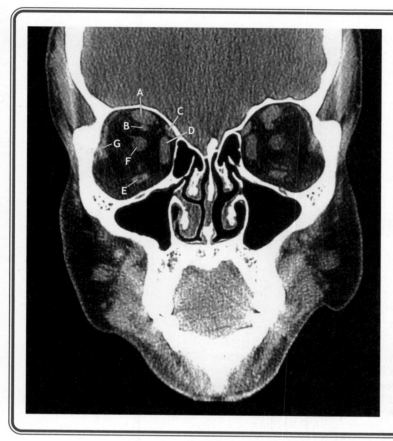

Figure 4–9. Coronal CT image of the orbit. **A:** Levator palpebrae superioris. **B:** Superior rectus. **C:** Superior oblique. **D:** Medial rectus. **E:** Inferior rectus. **F:** Optic nerve. **G:** Lateral rectus.

3. The **abducens nuclei lie beneath facial colliculi,** swellings in the floor of the fourth ventricle in the caudal pons (Figures 4–5A and 4–5B).
  a. The facial colliculus is formed by skeletal motor axons of the facial nerve that course around the abducens nucleus (see Figure 4–5A).
  b. The abducens nerve innervates the lateral rectus muscle, which acts to abduct the eye.
  c. The abducens neurons are situated in the PPRF, a center for horizontal conjugate gaze.

ABDUCENS NERVE LESIONS AND INTERNAL STRABISMUS

*A **lesion of the abducens nerve** may result in a **medially deviated eyeball (internal strabismus).***

4. The **hypoglossal nuclei** are located in the rostral part of the medulla; the hypoglossal nerve **innervates all of the muscles that act on the tongue, except the palatoglossus.**

HYPOGLOSSAL NERVE LESIONS AND TONGUE DEVIATION ON PROTRUSION

*A **hypoglossal nerve lesion** results in an **ipsilateral paralysis** of one half of the tongue and a **deviation of the tip of the tongue** on protrusion toward the side of the lesion.*

    B. An intermediate column of 4 nuclei contains **preganglionic parasympathetic neurons**; their axons exit the brainstem in CNs III, VII, IX, and X (see Figure 4–8).
        1. The **Edinger-Westphal nucleus** is in the rostral midbrain dorsal and lateral to the oculomotor nucleus.
           a. Preganglionic parasympathetic axons from the Edinger-Westphal nucleus exit the midbrain with the oculomotor nerve and synapse with postganglionic neurons in the ciliary ganglion.
           b. Axons from the ciliary ganglion supply the sphincter pupillae and ciliary muscles in the orbit.
           c. The parasympathetic axons of CN III form the motor limb of the light reflex and the motor limb of the accommodation (near response) reflex.

OCULOMOTOR NERVE LESIONS, A DILATED PUPIL, AND A LOSS OF THE NEAR RESPONSE

- *The **preganglionic parasympathetic axons** course in the peripheral part of CN III and are **subject to external compression before the skeletal motor fibers**.*
- *The initial signs of the **CN III compression** are a dilated pupil, loss of the light reflex, and loss of the near response on the side of the lesion.*

        2. The **superior salivatory nucleus** is in the caudal pons and gives rise to preganglionic parasympathetic axons in the facial nerve.
           a. Axons of the superior salivatory nucleus synapse with postganglionic neurons in the submandibular ganglion, which innervate the submandibular and sublingual glands.
           b. Axons of the superior salivatory nucleus synapse in the pterygopalatine ganglion, which innervates oral and nasal mucous glands and the lacrimal gland.

FACIAL NERVE LESIONS AND A DRY EYE

*A reliable diagnostic indicator of a **lesion of parasympathetic axons in CN VII** is a loss of **lacrimation** that results in a **dry eye on the side of the lesion** in affected patients.*

        3. The **inferior salivatory nucleus** is in the rostral medulla and gives rise to preganglionic parasympathetic axons in the glossopharyngeal nerve.
           a. Axons of the inferior salivatory nucleus synapse with postganglionic neurons in the otic ganglion.
           b. Postganglionic parasympathetic axons of the otic ganglion innervate the parotid gland.

GLOSSOPHARYNGEAL NERVE LESIONS AND A REDUCTION IN PAROTID SECRETIONS

- *A **lesion of CN IX** may result in a **reduction in parotid secretions**.*
- *Reduction in parotid secretions is difficult to use as a diagnostic indicator in patients with a CN IX lesion.*

4. The **dorsal motor nucleus of CN X** is located just lateral to the hypoglossal nucleus in the floor of the fourth ventricle; its preganglionic parasympathetic axons exit with the vagus nerve.

    a. Axons of the dorsal motor nucleus of CN X synapse with postganglionic parasympathetic axons in terminal ganglia situated in thoracic and abdominal viscera.

    b. Postganglionic parasympathetic axons of terminal ganglia act to slow heart rate, constrict bronchial smooth muscle, and promote peristalsis in gastrointestinal (GI) tract viscera.

C. A **lateral column of motor nuclei** contains lower motor neurons that innervate skeletal muscles derived from pharyngeal arches; their axons exit in CNs V, VII, IX, and X (see Figure 4–8).

    1. The **motor nucleus of V** is located in the rostral pons.

        a. The motor nucleus of V is adjacent to the point of entry and exit of all of the fibers of CN V.

        b. Axons of the motor nucleus innervate the 4 main muscles of mastication (masseter, temporalis, and medial and lateral pterygoids) and the mylohyoid muscle, anterior belly of the digastric muscle, tensor tympani muscle, and tensor veli palatini muscle.

TRIGEMINAL NERVE LESIONS AND DEVIATION OF THE MANDIBLE ON PROTRUSION

A *unilateral lesion of the motor fibers of CN V* may result in a deviation of the jaw on protrusion toward the side of the lesion.

    2. The **facial motor nucleus** is in the caudal pons; its axons loop around the abducens nucleus and form the facial colliculus in the floor of the fourth ventricle (Figures 4–5A and 4–5B).

        a. Lateral to the **genu,** facial motor axons join the sensory and parasympathetic fibers (of the nervus intermedius of CN VII) and exit ventrolaterally at the pontomedullary junction.

        b. Facial motor neurons innervate muscles of facial expression and the stapedius muscle, stylohyoid muscle, and posterior belly of the digastric muscle.

FACIAL NERVE LESIONS AND BELL'S PALSY

- *Lesions of the skeletal motor axons in the facial nerve* may result in a complete paralysis of muscles of facial expression ipsilateral to the side of the lesion and hyperacusis, a hypersensitivity to loud sounds caused by a weakness of the stapedius.
- *Patients have weakness in the ability to wrinkle the forehead, shut the eye, flare a nostril, and show their teeth on the side of the lesion.*

    3. The **nucleus ambiguus** contains lower motor neurons that exit the medulla in CNs IX and X (Figures 4–6A, 4–6B, and 4–8).

        a. CN IX innervates only the stylopharyngeus, a weakness of which is of no diagnostic value in evaluating the integrity of CN IX.

        b. CN X innervates all skeletal muscles of the palate except the tensor veli palatini, all the skeletal muscles of the pharynx except the stylopharyngeus, and all the skeletal muscles of the larynx.

c. Some neurons in the nucleus ambiguus give rise to preganglionic parasympathetic neurons, which contribute to the innervation of the heart.

VAGUS NERVE LESIONS AND WEAKNESS OF THE PALATE, PHARYNX, AND LARYNX

- *A **lesion of motor axons of CN X** that arise from the nucleus ambiguus may result in an ipsilateral **weakness of the soft palate** and a **nasal regurgitation of liquids**. The tip of the uvula may deviate away from the side of the lesion.*
- *A **weakness of pharyngeal muscles** may result in difficulty in swallowing (dysphagia), and a **weakness of laryngeal muscles** may result in hoarseness.*

4. The **accessory nerve** (CN XI) arises from an accessory nucleus in the cervical spinal cord from C1 to C5.
  a. The **axons of CN XI** pass through the foramen magnum and exit the skull through the jugular foramen with CNs IX and X.
  b. The accessory nerve **innervates the sternocleidomastoid and trapezius muscles**.

BRAINSTEM LESIONS AND THE ACCESSORY NERVE

- *Lesions of the medulla may affect fibers of CNs IX, X, or XII but not CN XI because the fibers of CN XI arise from the cervical spinal cord.*
- *CN XI is commonly lesioned outside the skull and results in a **weakness in the ability to laterally rotate the scapula** during abduction and a **weakness in the ability to elevate the scapula** (trapezius). Accessory nerve lesions may also result in a **weakness in the ability to turn the chin** to the side opposite the lesion (sternocleidomastoid).*

D. **Corticobulbar (bulb = brainstem) fibers** provide the upper motor neuron innervation to lower motor neurons in cranial nerve nuclei (Figures 4–2A, 4–2B, and 4–10).
  1. Corticobulbar fibers arise from motor cortex and synapse with lower motor neurons in the trigeminal motor nucleus, facial motor nucleus, nucleus ambiguus, hypoglossal nucleus, and accessory nucleus in the cervical spinal cord.
  2. The corticobulbar innervation of most cranial nerve lower motor neurons is predominantly bilateral.
  a. Lower motor neurons in the trigeminal motor nucleus, nucleus ambiguus, and hypoglossal nucleus are innervated by corticobulbar neurons in both the right motor cortex and the left motor cortex.
  b. In some individuals, in the hypoglossal nucleus and the nucleus ambiguus, a **contralateral corticobulbar innervation may predominate over an ipsilateral corticobulbar innervation**.

LESIONS OF THE HYPOGLOSSAL AND VAGUS NERVES VERSUS CORTICOBULBAR LESIONS TO THE HYPOGLOSSAL NUCLEUS AND NUCLEUS AMBIGUUS

- *In individuals with a **lesion of corticobulbar axons to the hypoglossal nucleus**, the **tongue muscles will not undergo fasciculations and atrophy** and may deviate **away** from the injured corticobulbar fibers. (Recall that in patients with a hypoglossal nerve lesion, the tongue deviates **toward** the side of the lesioned nerve.)*

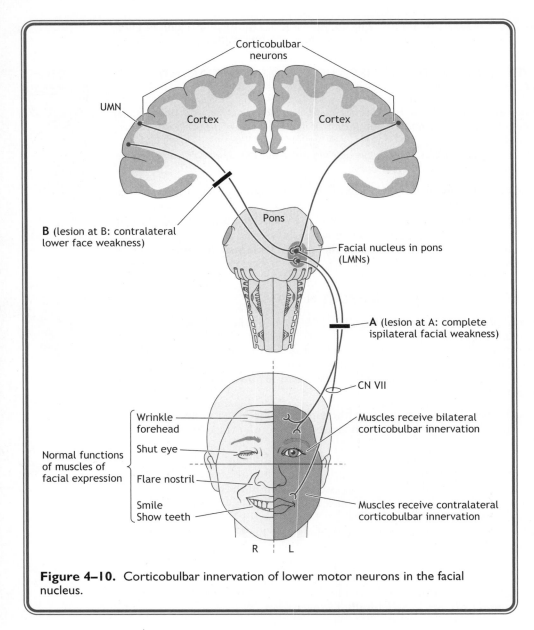

Figure 4–10. Corticobulbar innervation of lower motor neurons in the facial nucleus.

- In a lesion of corticobulbar axons to the nucleus ambiguus, the uvula may deviate **toward** the lesioned corticobulbar fibers. (Recall that in a vagus nerve lesion the uvula deviates **away** from the side of the lesioned nerve.)

   3. The corticobulbar innervation of the facial motor nucleus is bilateral for certain lower motor neurons and contralateral for other lower motor neurons (see Figure 4–10).

a. The corticobulbar innervation is bilateral to facial motor neurons that innervate muscles of the upper face, which act to wrinkle the forehead and shut the eye.

b. The corticobulbar innervation is exclusively contralateral to facial motor neurons that innervate muscles of the lower face, including those that act as sphincters and dilators of the nostrils and mouth.

LESIONS OF THE FACIAL NERVE VERSUS LESIONS OF CORTICOBULBAR FIBERS TO THE FACIAL MOTOR NUCLEUS AND THE SIDE OF THE DEFICITS

- *In patients with facial weakness, one may be able to differentiate between a lesion of CN VII and a lesion of the corticobulbar fibers to the facial motor nucleus. Patients with a **facial nerve lesion** (as in Bell's palsy) may have a **complete paralysis of muscles of facial expression ipsilateral** to the side of the lesioned nerve and have an **inability to wrinkle the forehead, shut the eye, and flare a nostril**, and a **drooping of the corner of the mouth**.*
- *A **unilateral corticobulbar lesion** may result in only a **lower face weakness**, as evidenced by a drooping of the corner of the mouth on the side of the face **contralateral** to the lesioned corticobulbar fibers. These patients will be able to wrinkle their forehead and shut their eyes and will have an intact blink reflex.*

4. **Corticobulbar fibers** also synapse with interneurons in conjugate gaze centers in the midbrain and pons that project to the oculomotor, trochlear, and abducens nuclei (see Chapter 7).

V. **Sensory nuclei of cranial nerves (see Figures 4–2 through 4–8) are in the lateral part of the brainstem and consist of the solitary nucleus, 3 trigeminal nuclei, and the vestibular and cochlear nuclei.**

A. The **solitary nucleus** is in the medulla (Figures 4–6A, 4–6B, and 4–8).

1. The rostral part of the solitary nucleus **receives taste fibers** conveyed into the CNS by the facial, glossopharyngeal, and vagus nerves.

a. CN VII conveys taste sensations from the anterior two thirds of the tongue and the soft palate.

b. CN IX conveys taste sensations from the posterior third of the tongue.

c. CN X conveys taste sensations from mucosa near the epiglottis.

d. The solitary nucleus relays taste to the ventral posterior medial nucleus of the thalamus, which projects to the somatosensory cortex.

2. The caudal part of the solitary nucleus is a **cardiorespiratory center**.

a. Neurons in the caudal part of the solitary nucleus **respond to chemoreceptors that monitor levels of carbon dioxide and oxygen in the blood, baroreceptors that monitor changes in blood pressure, and stretch receptors in the lungs**.

b. Chemoreceptors and baroreceptors in the carotid body and carotid sinus, respectively, are innervated by the glossopharyngeal and vagus nerves.

c. Chemoreceptors and baroreceptors in the aortic arch and stretch receptors in the lungs are innervated by CN X.

3. The solitary nucleus also **receives general sensations** conveyed into the medulla by CNs IX and X from the mucous membranes of most of the pharynx and the larynx.

RESPIRATORY FAILURE

- *A **bilateral lesion of the solitary nuclei** may cause respiratory failure*.
- *Respiratory failure may be preceded by **ataxic respiration** (irregular breathing)*.

GAG AND COUGH REFLEX

- *The **gag reflex** uses visceral sensory fibers in CN IX and skeletal motor fibers in CN X.*
- *The **cough reflex** uses both sensory and motor fibers in CN X.*

B. Three **trigeminal sensory nuclei respond to touch, pain, and temperature sensations** from the face, scalp, oral cavity, nasal cavity, and dura and to proprioception from muscles innervated by cranial nerves (Figures 4–11A and 4–11B).

1. Most sensory information that reaches the **trigeminal complex** enters the brainstem by way of primary sensory neurons in CN V, which have their cell bodies in the trigeminal ganglion.

2. **CNs VII, IX, and X** that convey general sensations from skin in and around the external auditory meatus also project to the trigeminal sensory nuclei.

3. The **principal sensory nucleus of V** is dorsolateral to the motor nucleus of V and is located in the rostral pons at the point of entry of fibers of the trigeminal nerve.

   a. The principal or main sensory nucleus of V receives central projections from primary sensory neurons mainly in the trigeminal ganglion that convey discriminative touch sensations.

   b. The axons of the principal sensory neurons cross and ascend bilaterally in the dorsal trigeminal or trigeminothalamic tracts and project to the ventral posterior medial nucleus of the thalamus.

   c. The axons of neurons in the ventral posterior medial nucleus of the thalamus project to the somatosensory cortex.

4. The **spinal nucleus of V** has 3 components that begin at the point of entry of the trigeminal nerve in the rostral pons and extend caudally through the length of the pons and medulla.

   a. The **oral** part of the spinal nucleus is an extension of the principal sensory nucleus in the caudal pons and also receives tactile inputs from the face, scalp, oral cavity, and nasal cavity.

   b. The **interpolar** part of the spinal nucleus is in the rostral medulla and responds to pain and temperature sensations from the tooth pulp.

   c. The **caudal** part of the spinal nucleus extends through the caudal medulla to upper cervical spinal cord levels and responds to pain and temperature sensations from the face, scalp, oral cavity, nasal cavity, and supratentorial dura.

   d. The spinal nucleus has a concentric "onion skin" topographic representation where skin and mucosa near the lips and nose are represented rostrally, and posterolateral areas of the face and scalp are represented more caudally (Figure 4–11A).

   e. The **axons arising from cells in all parts of the spinal nucleus of V** cross the brainstem diffusely to form the ventral trigeminal or trigeminothalamic tract.

   f. The **ventral trigeminothalamic tract** joins the medial lemniscus in the rostral pons, courses through the midbrain, and projects to the ventral posterior medial nucleus of the thalamus.

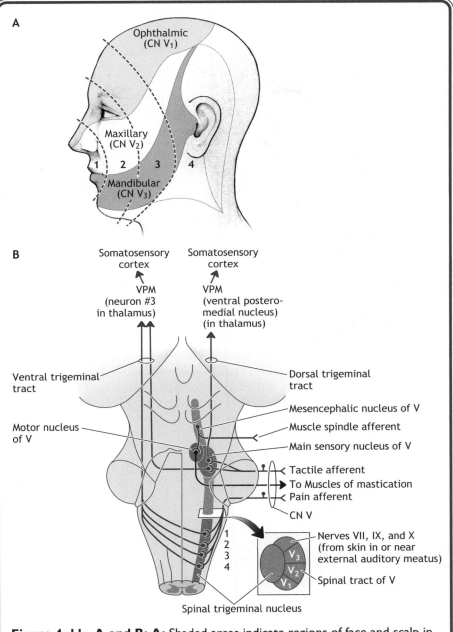

Figure 4–11. A and B: A: Shaded areas indicate regions of face and scalp innervated by branches of the three divisions of CN V. Dotted lines indicate concentric numbered "onion skin" regions emanating posteriorly from nose and mouth that have a rostral to caudal representation in the spinal nucleus of V in the brainstem. **B:** Fibers of CN V and neural components of trigeminal pathways. The numbers 1–4 correspond to the rostral caudal concentric representation of the face in Figure 4–11A. (**A:** Reproduced with permission from White, JS. *USMLE Road Map: Gross Anatomy*, 2006.)

5. The **mesencephalic nucleus of V** begins at the level of entry of CN V and extends into the midbrain lateral to the fourth ventricle and the cerebral aqueduct.

    a. Cells of the mesencephalic nucleus of V, like other primary sensory neurons, are derived from neural crest cells but are situated inside the midbrain instead of outside the CNS.

    b. The peripheral processes of mesencephalic neurons innervate stretch receptors in the muscles of mastication and in skeletal muscles innervated by other cranial nerves.

    c. The central processes of mesencephalic neurons project to the principal nucleus of V and to the motor nucleus of V as part of the jaw jerk reflex.

TRIGEMINAL NEURALGIA (TIC DOULOUREUX)

CLINICAL CORRELATION

- *Trigeminal neuralgia is characterized by **episodes of sharp, stabbing pain** radiating over the territory supplied by mucosal or cutaneous branches of the maxillary or mandibular divisions of the trigeminal nerve.*
- *The pain is frequently **triggered by moving the mandible, smiling, or yawning or by cutaneous or mucosal stimulation**.*

VI. The cochlear nerve (CN VIII) and the central auditory system process sound and participate in the localization of a sound source.

    A. The **external ear and middle ear** act to collect, amplify, and transmit sound vibrations outside the head to hair cells in the inner ear (Figures 4–12A and 4–12B).

        1. Sound vibrations represent **pressure fluctuations in air or bone** that vary in frequency and intensity.

        2. Frequency is perceived as the pitch of the sound and is measured in hertz.

            a. The **frequency range** for a child is 20–20,000 Hz; beginning at the age of 20, the ability to perceive high frequencies decreases at a rate of 200–300 Hz per year.

            b. The auditory system is most sensitive to frequencies between 2000 and 4000 Hz, a range that includes the frequencies of conversational speech.

        3. The **intensity of sound** is perceived as loudness and is expressed in decibels, (dB), which are based on a logarithmic scale.

            a. The **decibel scale** begins at a threshold of 0 dB and ranges up to 120 dB.

            b. A 120-dB sound has the intensity of a clap of thunder.

        4. The **external ear** (auricle and external auditory meatus) collects and transmits sound to the tympanic membrane, which vibrates in response to the sound stimulus.

        5. The **middle ear** contains 3 ossicles (malleus, incus, and stapes) that amplify the intensity of airborne auditory vibrations to overcome an impedance mismatch at the air-fluid interface between the middle ear and the inner ear (see Figure 4–12C).

            a. Sound is transmitted from the tympanic membrane by the malleus and the incus to the stapes, which is situated in the oval window.

            b. In the oval window, the **footplate of the stapes** is situated against the fluid environment of the inner ear that contains the receptor or hair cells.

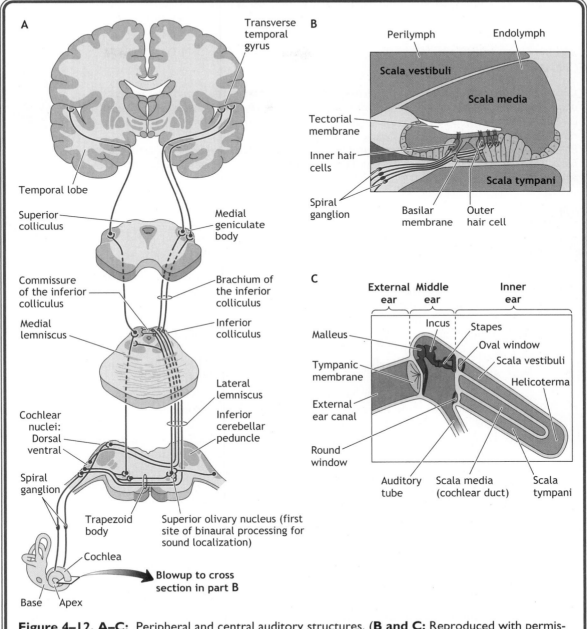

Figure 4–12. A–C: Peripheral and central auditory structures. (**B and C:** Reproduced with permission from Waxman, SG. *Clinical Neuroanatomy.* McGraw-Hill, 2003.)

c. The **impedance matching** by the middle ear is achieved mainly by an increase in force per unit area as a result of a 22:1 difference in the area of the tympanic membrane compared with the stapedial footplate.

CONDUCTIVE HEARING LOSSES

- *Conductive hearing losses result from **interference of sound transmission** through the external ear or middle ear. **Middle ear infections in children and otosclerosis in adults** are common causes of a conductive hearing loss. In conductive hearing loss, bone conduction is better than air conduction as a result of loss of the amplification provided by the middle ear.*
- *Patients with a conductive hearing loss will hear the vibrations better on the side of the defective external or middle ear because vibrations that reach the normal ear by both bone and air conduction interfere with each other, making the normal ear less sensitive.*

OTOSCLEROSIS

In patients with otosclerosis, the stapes becomes fused in the oval window; this results in a 20- to 30-dB loss in sensitivity, effectively reducing the loudness of normal conversation to that of a whisper.

    d. The **tensor tympani** (innervated by CN V) and **stapedius** (innervated by CN VII) **muscles** limit movement of the ossicles and protect the inner ear from high-intensity sounds; these muscles also contract to suppress sound transmission during one's own speech.

6. The **inner ear** consists of a series of interconnected fluid-filled sacs and channels that contain hair cells (see Figure 4–12B, C).
    a. The **auditory hair cells** are in an elongated cochlear duct in the **organ of Corti**.
    b. The hairs of all inner ear hair cells are **bathed in endolymph**, an extracellular fluid with an inorganic ionic composition (high concentration of K^+ ions) similar to that of an intracellular fluid.
    c. **Endolymph is produced by the stria vascularis**, which is in a wall of the cochlear duct; maintenance of the high K^+ ion concentration of endolymph is essential for the function of hair cells.

7. There are **2 populations of cochlear hair cells**.
    a. Both types of hair cell are situated on the flexible basilar membrane in the organ of Corti.
    b. A single row of inner hair cells forms the actual auditory receptor cells that are innervated by 95% of cochlear nerve fibers.
    c. There are 3 rows of outer hair cells, which are contractile and function to sharpen the frequency response characteristics of the cochlea.
    d. The outer hair cells are mainly innervated by (olivocochlear) efferent neurons from the brainstem and by only 5% of the cochlear nerve fibers.

8. The **inner hair cells and the basilar membrane** are tonotopically organized (see Figure 4–12B).
    a. At the **base of the cochlea**, high-frequency sounds cause maximum displacement of the basilar membrane and stimulation of hair cells.
    b. At the **apex of the cochlea**, low-frequency sounds cause maximum displacement of the basilar membrane and stimulation of hair cells.

B. At the **pontomedullary junction**, all cochlear nerve fibers bifurcate on entry into the brainstem and give rise to processes that synapse in both the ventral and dorsal cochlear nuclei. The spiral ganglion in the cochlea contains the cell bodies of the cochlear nerve fibers (see Figure 4–12A).

    1. The **ventral cochlear nuclei** project bilaterally to neurons in the superior olivary nuclei in the pons.

        a. The **superior olivary nuclei** are the first auditory nuclei to receive binaural inputs (Figures 4–5A and 4–5B).

        b. The **superior olivary and other central auditory nuclei** use differences in arrival time or intensity of binaural inputs to **localize the source of a sound**.

    2. The **lateral lemniscus** conveys auditory input from the cochlear nuclei and the superior olivary nuclei to the inferior colliculus in the midbrain.

    3. The **inferior colliculus** sends auditory information to the medial geniculate body (MGB) of the thalamus.

    4. Axons from the medial geniculate body project to the primary auditory cortex located on the posterior part of the transverse temporal gyrus.

        a. Primary auditory cortex corresponds to **Brodmann's areas** 41 and 42.

        b. Low-frequency sounds are represented in the rostral part of the primary auditory cortex, and high-frequency sounds are represented in the caudal part of the primary auditory cortex.

SENSORINEURAL HEARING LOSSES

- *Sensorineural hearing losses* may result from a loss of hair cells in the cochlea or from a lesion to the cochlear part of CN VIII or to any CNS auditory structure.
- At cochlear levels, there are **4 common causes of a sensorineural hearing loss**: trauma from high-intensity sound, infections, drugs, or presbycusis. **Presbycusis** is the most common cause of sensorineural hearing loss in the elderly and results from progressive high-frequency hair cell loss near the base of the cochlea.
- In patients with an **ipsilateral sensorineural hearing loss**, the lesion is most likely in the inner ear, CN VIII, or cochlear nuclei but not at higher levels of the central auditory system, which contain neural structures that receive inputs from both ears.
- A **lesion to CNS auditory structures in the brainstem** above the cochlear nuclei, in the thalamus, or in the cortex may result in a slight bilateral hearing loss and a decreased ability to localize a sound source.

VII. The vestibular nerve (CN VIII) (Figure 4–1A) innervates hair cells in several vestibular receptors.

    A. It innervates hair cells **in the ampullary crests** in the 3 semicircular ducts and hair cells **in the utricular and saccular maculae**.

        1. The **utricle and saccule** contain hair cells in maculae that respond to linear acceleration and detect positional changes in the head relative to gravity.

        2. The **semicircular ducts** contain hair cells in ampullary crests that respond to angular acceleration resulting from movements of the head.

        3. Vestibular hair cells are also innervated by axons of efferent neurons that have their neuron cell bodies inside the brainstem.

B. Central processes of vestibular nerve fibers enter at the pontomedullary junction and terminate in the **vestibular nuclei** and in the **flocculonodular lobe of the cerebellum**.

1. The vestibular nuclei extend from rostral medulla into the caudal pons; there are 4 main vestibular nuclei (superior, lateral, medial, and inferior) (see Figures 4–6A, 4–6B, and 4–8).

   a. The **medial and superior vestibular nuclei** receive vestibular nerve input from ampullary crests and are used in the vestibulo-ocular reflex (see later discussion).

   b. The **medial vestibular nucleus** is also used in conjugate eye movements.

   c. The **medial vestibulospinal tract** arises from the medial vestibular nucleus and coordinates neck and head movements in response to vestibular or visual stimuli.

   d. The **lateral nucleus** receives vestibular nerve input from ampullary crests, maculae, and the vestibulocerebellum.

   e. The **lateral vestibulospinal tract** arises from the lateral vestibular nucleus and facilitates extensor muscles.

   f. The **inferior nucleus** receives vestibular nerve input mainly from utricular and saccular maculae and integrates vestibular and motor signals.

LESIONS OF THE VESTIBULAR NERVE OR NUCLEI AND VERTIGO

CLINICAL CORRELATION

- *Patients with a **vestibular lesion** may have **vertigo**, an abnormal perception of rotation, which may involve either the subject or the external space. The vertigo is usually severe in peripheral vestibular lesions and mild in central CNS lesions.*
- *Patients with **Ménière's disease** may have **abrupt, recurrent attacks of vertigo** lasting minutes to hours. The vertigo may be accompanied by **tinnitus** (ringing) and an **ipsilateral sensorineural hearing loss. Nausea, vomiting, and a sensation of fullness or pressure in the ear** also are common during the acute episode. The disease usually occurs in middle age and **results from distention of the endolymph-filled spaces in the cochlear and vestibular parts of the labyrinth**.*

2. **Axons of the vestibular nuclei** join the MLF (see Figures 4–3 to 4–6).

   a. Axons in the MLF interconnect the vestibular nuclei with centers for conjugate gaze and with the oculomotor, trochlear, and abducens nuclei.

   b. The MLF also contains axons of the medial vestibulospinal tract.

C. The **vestibulo-ocular reflex** generates conjugate eye movements to enable the eyes to remain focused on a stationary target when the head moves (see Figures 4–13 and 4–14).

1. The vestibulo-ocular reflex uses the vestibular labyrinth, vestibular nerve, vestibular nuclei, MLF, and neurons in the abducens and oculomotor nuclei.

2. The vestibulo-ocular reflex results in reflex conjugate eye movements in a direction opposite that of head turning.

   a. When the head rotates horizontally to the left, **endolymph flows to the right, stimulating hair cells in the ampullary crest** in the horizontal semicircular duct in the left labyrinth. (Figure 4–13, #1).

   b. **Hair cells in the left horizontal semicircular duct** stimulate the left vestibular nerve and the left vestibular nuclear complex. (Figure 4–13, #2, #3).

   c. The **left vestibular nuclear complex** sends axons by way of the medial longitudinal fasciculus to the left oculomotor nucleus and to the right abducens nucleus. (Figure 4–13, #4).

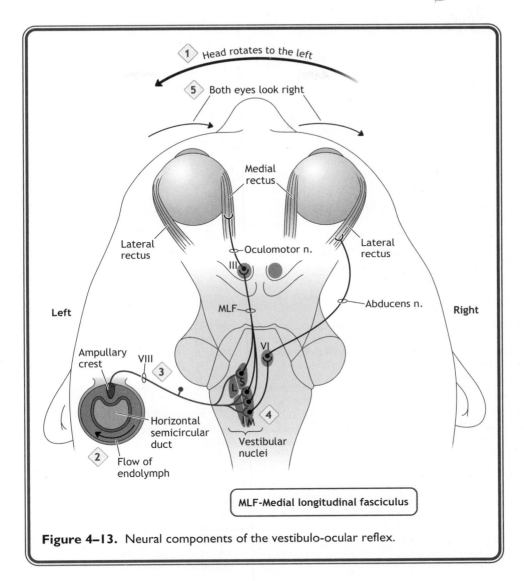

Figure 4–13. Neural components of the vestibulo-ocular reflex.

MLF-Medial longitudinal fasciculus

d. The **left oculomotor nerve** stimulates the left medial rectus, which adducts the left eye, and the right abducens nerve stimulates the right lateral rectus, which abducts the right eye.

e. The **net effect of horizontal rotation of the head to the left is horizontal conjugate gaze to the right**. (Figure 4–13, #5).

LESIONS OF VESTIBULAR STRUCTURES AND A VESTIBULAR NYSTAGMUS

• **Nystagmus** is a slow, rhythmic oscillation of the eyes to one side followed by a fast reflex movement of the eyes in the opposite direction. Nystagmus is usually horizontal, although rotatory or vertical nystagmus may also occur.

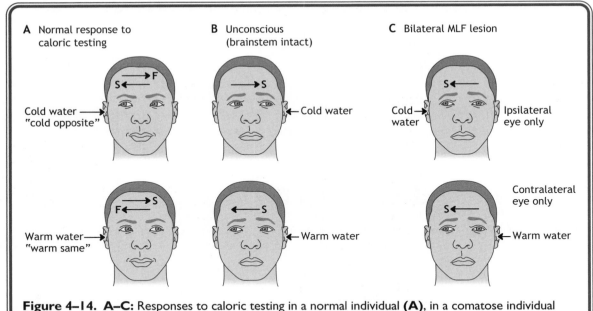

Figure 4–14. A–C: Responses to caloric testing in a normal individual **(A)**, in a comatose individual with a cortex lesion, but an intact brainstem **(B)**, and in a comatose individual with a bilateral MLF lesion **(C)**. S, slow phase; F, fast phase. (Reproduced with permission from Michael-Titus, Revest, and Shortland. *The Nervous System,* Elsevier, 2007.)

- *Nystagmus is defined by the direction of the fast, or corrective, phase. Nystagmus may **result from a lesion of the labyrinth, vestibular nerve, or nuclei**. Nystagmus also may result from a **metabolic disease or a lesion of the cerebellum, visual system (usually reflex centers and their connections), or cerebral cortex**.*
- *In a pathological nystagmus involving the vestibular nerve, there is an initial slow deviation of the eyes toward the side of the lesion in response to the pathology. A fast or corrective phase occurs in the opposite direction that is generated by the cerebral cortex. In patients with a vestibular-evoked nystagmus, the fast or corrective phase is **away** from the side of the lesion. A peripheral nystagmus is usually accompanied by vertigo. In a central lesion of the vestibular nuclei or cerebellum, the nystagmus may be bidirectional and may occur without vertigo.*

    D. The **caloric test** evaluates the vestibulo-ocular reflex by introducing warm or cool water into an external auditory meatus (Figure 4–14A).

       1. Warm water introduced into the external auditory meatus warms the endolymph and stimulates a horizontal semicircular duct, causing the eyes to move slowly away from the irrigated meatus.

       2. Normally, the eyes will move quickly back to the irrigated side, producing a fast phase of a caloric-evoked nystagmus toward the same side of the warm water stimulus.

       3. Introduction of cool water into an external auditory meatus cools the endolymph and mimics a lesion.

          a. The horizontal semicircular duct is inhibited on the cool water side.

          b. The opposite vestibular complex causes the eyes to move slowly toward the irrigated meatus.

4. A fast phase of the nystagmus moves the eyes quickly away from the side of the cool water stimulus.

CALORIC TESTING AND "COWS"

*The direction of the fast phase of vestibular-evoked nystagmus in a caloric test toward the warm water side and away from the cool water side is summarized by the **mnemonic COWS** (cool, opposite, warm, same).*

CALORIC TESTING AND THE COMATOSE PATIENT (FIGURE 4–14B)

In a comatose patient with a nonfunctioning cerebral cortex, but an intact brainstem, caloric testing will only generate a slow deviation of both eyes, but not a fast or corrective phase. In a comatose patient with a bilateral MLF lesion above the abducens nucleus, caloric testing will only result in a slow deviation of the abducting eye because adduction requires that the MLF be intact (Figure 4–14C).

In a patient with a bilateral lesion to the vestibular nuclei, there will be no eye movement at all in response to caloric testing.

VIII. Branches of the vertebral and basilar arteries in the posterior circulation supply the brainstem and cerebellum (Figure 4–15).

A. The **vertebral arteries** course on the ventral surface of the medulla and join to form the basilar artery at the pontomedullary junction.

1. Each vertebral artery gives rise to a root of the anterior spinal artery that supplies the ventral and medial aspects of the medulla (see Figure 4–15A).

2. Each vertebral artery gives rise to a posterior inferior cerebellar artery (PICA), which supplies the lateral medulla and the cerebellum (see Figure 4–15A).

3. Direct branches of the vertebral artery supply an intermediate region of the medulla.

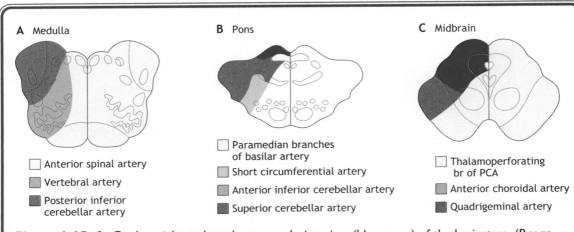

Figure 4–15. A–C: Arterial supply and common lesion sites (blue areas) of the brainstem. (Reproduced with permission from Michael-Titus, Revest, and Shortland. *The Nervous System,* Elsevier, 2007.)

B. The **basilar artery gives rise to several paramedian and short circumferential branches**, which supply the medial aspect of the pons and midbrain.

    1. The basilar artery gives rise to the **anterior inferior cerebellar arteries**, which supply the lateral aspect of the caudal pons and overlying cerebellum (see Figure 4–15B).

    2. It also gives rise to the **labyrinthine artery**, which supplies the inner ear.

    3. The basilar artery gives rise to the **superior cerebellar arteries**, which supply the superior cerebellar peduncle and the overlying cerebellum.

    4. The basilar artery ends at the rostral end of the pons by branching sharply into a pair of posterior cerebral arteries.

C. Deep branches of the proximal part of the posterior cerebral arteries (PCA) supply the midbrain (Figure 4–15C)

    1. Paramedian or thalamoperforating branches of the PCA supply the medial midbrain.

    2. The quadrigeminal artery that also arises from the PCA near the posterior communicating artery supplies the dorsolateral midbrain.

IX. Most brainstem syndromes are caused by a stroke involving a branch of one of the vertebral arteries, a branch of the basilar artery, or a branch of a posterior cerebral artery.

A. The most common **brainstem lesions** combine a lesion of 1 or more ascending or descending tracts with a lesion of 1 or more cranial nerves.

    1. A **lesion to a descending or ascending tract** (corticospinal, medial lemniscus, spinothalamic tract) results in deficits that are contralateral and below the lesion.

        a. Patients may have a contralateral spastic hemiparesis in the limbs (corticospinal lesion), contralateral loss of touch vibration and pressure sensations from the limbs and trunk (medial lemniscus lesion), or contralateral loss of pain and temperature sensations from the limbs and body (spinothalamic tract).

        b. The only **ipsilateral "long tract sign"** seen in a brainstem lesion is caused by a lesion of the descending hypothalamic fibers that results in Horner's syndrome ipsilateral to the side of the lesion.

    2. A **lesion to either sensory or motor axons in a cranial nerve** localizes the lesion to a specific part of the midbrain, pons, or medulla and results in signs or symptoms that are ipsilateral and at the level of the lesion.

        a. A lesion of the **oculomotor or trochlear nerve** localizes the lesion to the midbrain.

        b. A lesion of the **trigeminal nerve** localizes the lesion to the rostral pons.

        c. A lesion of the **abducens, facial, or vestibulocochlear nerve** localizes the lesion to the caudal pons.

        d. A lesion of the **glossopharyngeal, vagus, or hypoglossal nerve** localizes the lesion to the medulla.

BRAINSTEM LESION DEFICITS

- *Deficits associated with brainstem lesions commonly result in signs or symptoms that begin with the letter "d."*
- *Deficits include diplopia; a deviated eye, tongue, uvula, or mandible; dysphagia; dysphonia; a drooping eyelid; or a droopy face.*

B. Medial medullary syndrome most commonly **results from an occlusion of a vertebral artery or the anterior spinal artery** (see Figure 4–15A).

   1. Patients with medial medullary syndrome may have **2 long tract signs** caused by a lesion of the medial lemniscus and the corticospinal tract combined with a lesion of the hypoglossal nerve that localizes the lesion to the medial medulla.

   a. These patients have a **contralateral spastic hemiparesis** and a **contralateral loss of touch vibration and pressure in the limbs and trunk**.

   b. Lesions of the hypoglossal nerve result in **dysarthria**, caused by paralysis and atrophy of the tongue ipsilateral to the lesion, and a **deviation of the tongue** toward the side of the lesion on protrusion.

   2. These patients have an "**inferior alternating hemiplegia**," in which there is spastic weakness in the limbs on the side contralateral to the lesion and tongue weakness ipsilateral to the lesion.

C. Lateral medullary syndrome (Wallenberg's syndrome) is most commonly **caused by an occlusion of a vertebral artery or a PICA** (see Figure 4–15A).

   1. The **long tracts affected are the spinothalamic tract and the descending hypothalamic fibers**.

   a. Spinothalamic tract lesions result in a loss of pain and temperature sensations in the limbs and body contralateral to the lesion.

   b. A lesion of descending hypothalamic fibers results in Horner's syndrome with miosis, ptosis, and anhidrosis ipsilateral to the lesion.

   2. The cranial nerves that may be affected by the lesion and localize the lesion to the lateral medulla are **the vestibular or the cochlear nuclei, the glossopharyngeal nerve, and the vagus nerve**.

   a. A lesion of the vestibular nuclei may result in vertigo, nausea and vomiting, and a vestibular nystagmus, in which the fast phase will be away from the side of the lesion.

   b. A lesion of the cochlear nucleus results in an ipsilateral sensorineural hearing loss.

   c. A lesion of the vagus nerve as it exits the medulla is the most common localizing cranial nerve sign of lateral medullary syndrome.

   d. Vagus nerve lesions result in dysphagia (difficulty in swallowing) or hoarseness, the palate may droop on the affected side, and the uvula may deviate away from the side of the lesion.

   e. A lesion of a glossopharyngeal nerve may result in a diminished or absent gag reflex.

   3. The spinal tract and spinal nucleus of the trigeminal nerve may be lesioned.

   a. There may be loss of pain and temperature sensations from half of the face and scalp ipsilateral to the side of the lesion.

   b. Touch sensations from the face and the corneal blink reflex will be intact.

   c. Patients have a dissociated loss of pain and temperature; pain and temperature sensations are lost from the face and scalp **ipsilateral** to the lesion but are lost from the **contralateral** limbs and trunk.

   4. The solitary nucleus may be affected.

   a. Patients may have an alteration or loss of taste from the tongue ipsilateral to the side of the lesion.

   b. The facial and glossopharyngeal nerves convey sensations of taste into the solitary nucleus from the anterior two thirds and posterior third of the tongue, respectively.

D. Medial pontine syndrome may result from an occlusion of paramedian branches of the basilar artery (see Figure 4–15B).
1. Medial pontine syndrome **most commonly affects the corticospinal tract and the exiting fibers of the abducens nerve**.
2. Patients have a **spastic hemiparesis** contralateral to the lesion and an **internal strabismus** ipsilateral to the lesion.
3. The medial lemniscus may be affected if the lesion is deeper into the pons, and the facial nerve may be affected if the lesion extends laterally.
4. Patients may have a loss of touch, vibration, and pressure sensations in the limbs and body contralateral to the lesion and facial weakness, a dry eye, and alteration of taste from the anterior 2/3 of the tongue ipsilateral to the lesion.
5. The long tract signs are the same as those seen in medial medullary syndrome, but the lesion to the abducens nerve and the facial nerve localizes the lesion to the caudal pons.

E. Lateral pontine syndrome results in lesions of the caudal or rostral pons.
1. **Lesions of the dorsolateral pons** may result from an **occlusion of the anterior inferior cerebellar artery (caudal pons) or circumferential branches of the basilar artery** (see Figure 4–15B).
2. The **long tracts affected** by a lesion in the lateral and caudal pons are the same as in lateral medullary syndrome, **the spinothalamic tract, and the descending hypothalamic fibers**.
3. The **facial and vestibulocochlear nerves or nuclei** may be affected in a lateral pontine syndrome of the caudal pons.
   a. A lesion of the vestibular nuclei may result in nausea and vomiting and a vestibular nystagmus, in which the fast phase will be away from the side of the lesion.
   b. A lesion of the cochlear nucleus may result in a sensorineural hearing loss ipsilateral to the lesion.
4. **Fibers of the trigeminal nerve** may be affected in a lateral pontine syndrome of the rostral pons.
   a. Patients may have complete anesthesia of the face and scalp and weakness in muscles of mastication ipsilateral to the lesion.
   b. The tip of the jaw deviates toward the side of the lesion on protrusion.

F. Pontocerebellar angle syndrome most commonly results from a schwannoma.
1. Schwannomas develop in Schwann cells of the vestibulocochlear nerve just outside the brainstem (see Figure 4–1A).
   a. Initially, a patient with a vestibulocochlear schwannoma may have a combination of tinnitus (ringing in the ear) and difficulty maintaining balance.
   b. As the neoplasm enlarges, the patient may have a sensorineural hearing loss combined with dizziness and vertigo.
   c. The adjacent facial and trigeminal nerves may be compressed, resulting in facial and jaw weakness ipsilateral to the neoplasm.
   d. If the neoplasm compresses the middle or inferior cerebellar peduncle, the patient may have an ataxic gait with a tendency to fall toward the side of the neoplasm.
2. Patients with a vestibulocochlear schwannoma will not have long tract signs because the lesion is outside the brainstem.

G. Medial midbrain (Weber's) syndrome commonly results from an occlusion of thalamoperforating branches of the posterior cerebral artery (see Figure 4–15C).

1. It results in **lesions of the corticospinal and corticobulbar tracts and the oculomotor nerve**.

   a. A corticospinal tract lesion results in a spastic hemiparesis contralateral to the lesion.

   b. A lesion of an oculomotor nerve results in a dilated pupil, ptosis, external strabismus, and loss of the near response and pupillary light reflexes ipsilateral to the side of the lesion.

   c. A lesion of the corticobulbar fibers results in lower face weakness (drooping of the corner of the mouth) contralateral to the lesion. The patient will be able to shut the eye (blink reflex is intact) and wrinkle the forehead. The tongue may deviate away from the lesioned fibers, and the uvula may deviate toward the lesioned fibers.

2. These patients have a "**superior alternating hemiplegia,**" in which there is spastic weakness in the limbs on the side contralateral to the lesion and an external strabismus ipsilateral to the lesion.

H. Parinaud's syndrome most commonly results from compression by a pineal tumor.

1. A **pineal tumor compresses the pretectal area of the midbrain and the superior colliculi** (see Figure 4–1BG).

   a. **Compression of the pretectal area** may result in bilateral pupillary reflex abnormalities (e.g., slightly dilated and fixed pupils, which may show an impaired light and accommodation reaction).

   b. **Compression of the rostral interstitial nucleus** adjacent to the superior colliculus (see Chapter 7) may result in a paralysis of vertical gaze, and there may be a downward deviation of both eyes (a "**setting-sun**" **sign**).

2. A **pineal tumor may also compress the cerebral aqueduct in the midbrain**, resulting in an elevation of intracranial pressure and a noncommunicating hydrocephalus.

CLINICAL PROBLEMS

1. A patient has hoarseness and difficulty swallowing, loss of pain and temperature sensations from the body contralateral to the lesion and from the face ipsilateral to the lesion, and Horner's syndrome. A single lesion that accounts for all of the signs or symptoms is in:

   A. The medial medulla

   B. The lateral medulla

   C. The caudal pons

   D. The rostral pons

   E. The midbrain

   F. A cranial nerve or cranial nerves outside of the brainstem

2. A patient cannot wrinkle his forehead or shut his eye on the side of the lesion. The affected eye is also dry and red, and the patient complains of being sensitive to loud sounds. An internal strabismus is present on the side of the facial weakness. The patient also has weak upper and lower limbs and elevated muscle stretch reflexes contralateral to the facial weakness. A single lesion that accounts for all of the signs or symptoms is in:

   A. The medial medulla

   B. The lateral medulla

   C. The caudal pons

   D. The rostral pons

   E. The midbrain

   F. A cranial nerve or cranial nerves outside of the brainstem

3. A patient has a dilated pupil, a laterally deviated eye, loss of the near response on the side of the lesion, and weak upper and lower limbs contralateral to the side of the lesion. A single lesion that accounts for all of the signs or symptoms is in:

   A. The medial medulla

   B. The lateral medulla

   C. The caudal pons

   D. The rostral pons

   E. The midbrain

   F. A cranial nerve or cranial nerves outside of the brainstem

4. A patient cannot wrinkle her forehead or smile on the right and has sensorineural hearing loss and weak jaw muscles on the right. The patient notes that the hearing loss has been progressive over the years, but the facial weakness and jaw weakness were evident only recently. A single lesion that accounts for all the signs or symptoms is in:

   A. The medial medulla

   B. The lateral medulla

   C. The caudal pons

   D. The rostral pons

   E. The midbrain

   F. A cranial nerve or cranial nerves outside of the brainstem

5. A patient has a loss of vibratory sense from the left side of the body and a spastic hemiparesis on the left, and the tongue deviates toward the right on protrusion. What blood vessel may have been involved in a stroke?

   A. Posterior inferior cerebellar artery

   B. Basilar artery

   C. Anterior spinal artery

   D. Paramedian branches of the basilar artery

   E. Deep branches of a posterior cerebral artery

6. A patient has nasal regurgitation of liquids during swallowing and nasal speech, miosis and ptosis of the right pupil and eyelid, respectively, and a loss of pain and temperature sensations from the body opposite the ocular signs. The patient notes that food tastes funny and that the right side of the face is dry. What blood vessel may have been involved in a stroke?

   A. Posterior inferior cerebellar artery

   B. Basilar artery

   C. Anterior spinal artery

   D. Paramedian branches of the basilar artery

   E. Deep branches of a posterior cerebral artery

7. A patient has a laterally deviated left eye, a ptosis of the left eyelid, and a dilated left pupil. The patient can shut the affected eye but cannot prevent saliva from dripping from the corner of the mouth on the right. There are elevated muscle stretch reflexes and weakness in the right upper and lower limbs. What blood vessel may have been involved in a stroke?

   A. Posterior inferior cerebellar artery

   B. Basilar artery

   C. Anterior inferior cerebellar artery

   D. Paramedian branches of the basilar artery

   E. Deep (thalamoperforating) branches of a posterior cerebral artery

8. A patient has numbness of the face and scalp on the right, burns the anterior part of the tongue but cannot feel the stimulus, and has weakness in chewing on the right. When the left cornea is stimulated both eyes blink, but when the right cornea is stimulated neither eye blinks. Localize the probable lesion site.

   A. Trigeminal nerve

   B. Spinal nucleus of V

   C. Principal sensory nucleus of V

   D. Ventral trigeminal tract

   E. Mesencephalic nucleus of V

9. A 49-year-old secretary with a history of high blood pressure experienced a sudden onset of dizziness, nausea, and vomiting. She was brought to the emergency room, where a neurological exam revealed:

   • A horizontal nystagmus

   • Dysphagia and hoarseness

   • Absent gag reflex on the left

   • Alteration of taste from the tongue

   • Analgesia and thermal anesthesia on the left side of the face

   • Analgesia and thermal anesthesia on the right side of the body

   • Horner's syndrome

   • Significant hearing loss on the left compared with the right

The dysphagia and hoarseness in this patient may be due to a lesion in the:

A. Dorsal motor nucleus of X

B. Nucleus solitarius

C. Nucleus ambiguus

D. Inferior salivatory nucleus

E. Superior salivatory nucleus

10. The analgesia and thermal anesthesia on the left side of the face in this case most likely resulted from a lesion of:

A. The trigeminal nerve

B. The mesencephalic nucleus of V

C. The principal nucleus of V

D. The spinal tract of V

E. The trigeminal ganglion

11. What else might be observed in the patient other than the signs and symptoms noted previously?

A. The uvula may deviate to the left.

B. Sensations of touch might also be altered in the anterior two thirds of the tongue.

C. The pupil on the left will be dilated compared with the pupil on the right.

D. The horizontal nystagmus will have a quick component to the right.

E. Retrograde changes might be evident in neurons in the dorsal horn of the spinal cord on the left.

12. Your elderly patient has presbycusis that is more evident on the right than the left. What might you correctly conclude?

A. The patient has hair cell degeneration at the base of the cochlea.

B. The patient has excessive wax buildup in the external auditory meatus.

C. The patient has hyperacusis.

D. The patient has otosclerosis.

E. Bone conduction will be better than air conduction on the left.

13. A complete destructive lesion of the facial nerve just as it emerges from the brainstem will result in retrograde chromatolysis in which of the following nuclei?

A. Nucleus ambiguus

B. Inferior salivatory nucleus

C. Superior salivatory nucleus

D. Ventral cochlear nucleus

E. Solitary nucleus

14. During a neurological evaluation, you note that when you stimulate your patient's right cornea with a wisp of cotton, both eyes blink, but when you stimulate the left

cornea, there is no response. Your patient may also have which of the following signs or symptoms?

A. A dry eye

B. Altered sensations of taste from the tongue

C. A dilated pupil on the left

D. Altered sensation in the skin of the forehead

E. A drooping of the corner of the mouth on the left

15. By placing warm water in the patient's left external auditory meatus, under normal circumstances you would expect:

A. A nystagmus with a quick component to the right

B. Both eyes to drift slowly to the left

C. The left eye to look to the left

D. A nystagmus with a quick component to the left

E. Both eyes to look superiorly

16. A transverse section through the brainstem contains the solitary nucleus. What other structure might you expect to see in the same section?

A. Principal (chief) sensory nucleus of V

B. Facial motor nucleus

C. Spinal nucleus of V

D. Abducens nucleus

E. Trochlear nucleus

MATCHING PROBLEMS

Questions 17–42: Clinical features match

Choose from A–U the one most closely associated with the clinical deficit.

Choices (each choice may be used once, more than once, or not at all):

A. Facial motor nucleus

B. Solitary nucleus

C. Inferior salivatory nucleus

D. Motor nucleus of V

E. Edinger-Westphal nucleus

F. Abducens nucleus

G. Trochlear nucleus

H. Spinal nucleus of V

I. Mesencephalic nucleus of V

J. Dorsal motor nucleus of X

K. Superior salivatory nucleus

    L. Nucleus ambiguus

    M. Principal sensory nucleus of V

    N. Oculomotor nucleus

    O. Hypoglossal nucleus

    P. Trigeminal ganglion

    Q. Descending hypothalamic fibers

    R. Corticospinal fibers

    S. Corticobulbar fibers

    T. Anterolateral system

    U. Medial lemniscus

17. Internal strabismus

18. Tongue deviates upon protrusion

19. Uvula deviates during swallowing

20. Decreased output of parotid gland

21. Ptosis and constricted pupil

22. Diplopia and ptosis

23. Hoarseness

24. Loss of motor limb of blink reflex

25. Constricted pupil, ptosis, dry face

26. Loss of all facial sensation

27. Laterally deviated eye

28. Loss of motor limb of gag reflex

29. Inability to depress adducted eye

30. Saliva drips from corner of mouth; blink reflex intact bilaterally

31. Jaw deviates upon protrusion

32. Loss of pain and temperature sensations from face

33. No accommodation

34. Eye dry and red

35. Loss of ability to adduct an eye

36. Altered taste from posterior third of tongue

37. Disruption of sensory limb of jaw-jerk reflex

38. Cannot shut an eye

39. Loss of vibratory sensations from upper limb

40. Loss of taste from anterior two thirds of tongue

41. Hyperacusis

42. Delayed gastric emptying

Questions 43–56: Lesions match

Choose a location in A–F that mostly likely is a single lesion site that accounts for all the symptoms.

Choices: (each choice may be used once, more than once, or not at all):

    A. Medial medulla

    B. Lateral medulla

    C. Caudal pons

    D. Rostral pons

    E. Midbrain

    F. Lesion affects a cranial nerve or cranial nerves outside the brain stem

43. A patient has hoarseness and difficulty swallowing, a loss of pain and temperature sensations from the body contralateral to the lesion, and a loss of pain and temperature sensations from the face ipsilateral to the lesion.

44. A patient has nystagmus with a fast component away from the side of the lesion, Horner's syndrome on the side of the lesion, no gag reflex with hoarseness, and altered protopathic sensations from the face.

45. A patient cannot shut or abduct an eye, cannot wrinkle their forehead, and has spastic weakness contralateral to the facial and ocular signs.

46. A patient has a dilated pupil, a laterally deviated eye, a loss of the near response on the side of a lesion, and weak upper and lower limbs contralateral to the side of the lesion.

47. A patient cannot wrinkle their forehead or smile and has a sensorineural hearing loss and weak jaw muscles on the same side as the facial weakness.

48. A patient has a loss of vibratory sense from the limbs and body and a tongue that deviates on protrusion away from the side of the sensory loss.

49. A patient has nasal regurgitation of liquids during swallowing and nasal speech, miosis and a dry face, and a loss of pain and temperature sensations from the body contralateral to dry face.

50. A patient has atrophy and fasciculations of muscles on one side of the tongue, a loss of vibratory sensations, and weakness in the limbs and body contralateral to the side of the atrophied tongue.

51. A patient has an intact blink reflex, but has a laterally deviated eye, a ptosis on the same side, and a drooping of the corner of the mouth and a Babinski sign contralateral to the ocular signs.

52. A patient has complete anesthesia of the face and scalp, complete weakness of muscles of facial expression, and a sensorineural hearing loss all on the same side of a lesion.

53. A patient has a medially deviated eye; the patient cannot shut the affected eye, and there are elevated muscle stretch reflexes in the limbs contralateral to side with the ocular signs.

54. A patient has a complete anesthesia of the face and scalp and weakness in chewing, with weak upper and lower limbs contralateral to facial signs.

55. A patient has a lesion of neurons in the solitary nucleus, nucleus ambiguus, and in the spinal nucleus of V.

56. Patient has an internal strabismus, a dry eye, hyperacusis, and facial weakness on the side of a lesion, combined with a hearing loss and spastic weakness of the contralateral limbs.

57. Patient cannot keep saliva from dripping out of the corner of the mouth, has a normal blink reflex, but elevated stretch reflexes in weak limbs, and has a ptosis and a laterally deviated eye on the side opposite the weak limbs.

ANSWERS

1. The answer is B. The patient has lateral medullary syndrome. The long tracts affected in this syndrome are the spinothalamic tract and the descending hypothalamic fibers. The cranial nerves that may be affected by the lesion and localize the lesion to the lateral medulla are the vestibular or the cochlear nuclei, the glossopharyngeal nerve, and the vagus nerve. The spinal tract and spinal nucleus of the trigeminal nerve may be lesioned in lateral medullary syndrome.

2. The answer is C. The patient has a form of medial pontine syndrome that has affected the corticospinal tract, the exiting fibers of the abducens nerve, and the facial nerve.

3. The answer is E. The patient has medial midbrain syndrome. Medial midbrain syndrome results in a lesion of the corticospinal and corticobulbar tracts and the oculomotor nerve.

4. The answer is F. The patient has pontocerebellar angle syndrome, which most commonly results from a schwannoma that develops in Schwann cells of the vestibulocochlear nerve just outside the brainstem. Initially, a patient with a vestibulocochlear schwannoma may have a sensorineural hearing loss combined with dizziness and vertigo. The adjacent facial and trigeminal nerves may be compressed, resulting in facial and jaw weakness ipsilateral to the neoplasm. This patient had no long tract signs, indicating that the lesion is not inside the brainstem.

5. The answer is C. The patient has medial medullary syndrome, with 2 long tract signs (the medial lemniscus and the corticospinal tract), combined with a lesion of the hypoglossal nerve that localizes the lesion to the medial medulla.

6. The answer is A. The patient has lateral medullary syndrome. A lesion of descending hypothalamic fibers results in Horner's syndrome with miosis, ptosis, and anhidrosis ipsilateral to the lesion. Spinothalamic tract lesions result in a loss of pain and temperature sensations in the limbs and body contralateral to the lesion. The cranial nerves that may be affected by the lesion and localize the lesion to the lateral medulla are the vestibular or the cochlear nuclei, the glossopharyngeal nerve, and the vagus nerve.

7. The answer is E. The patient has medial midbrain syndrome. Medial midbrain syndrome results in a lesion of the corticospinal and corticobulbar tracts and the oculomotor nerve. A lesion of the corticobulbar fibers results in a contralateral lower face weakness.

8. The answer is A. The patient has a complete lesion affecting all the sensory and motor fibers of V.

9. The answer is C. The nucleus ambiguus contains lower motor neurons that exit the medulla in CN X and innervate all skeletal muscles of the palate except the tensor veli palatini, all the skeletal muscles of the pharynx except the stylopharyngeus, and all of the skeletal muscles of the larynx.

10. The answer is D. The analgesia and thermal anesthesia most likely resulted from a lesion of the spinal tract of V because the patient's loss is limited to pain and temperature on the side of the lesion.

11. The answer is D. The horizontal nystagmus will have a quick component to the right; the fast phase of a vestibular evoked nystagmus is away from the side of the lesion (the left). In this patient, the uvula may deviate to the right. Sensations of touch will not be altered in the anterior two thirds of the tongue. The pupil on the left will be constricted (Horner's syndrome) compared with the pupil on the right. Retrograde changes might be evident in neurons in the dorsal horn of the spinal cord on the right (left spinothalamic tract).

12. The answer is A. Hair cell degeneration at the base of the cochlea is the most common cause of presbycusis, a sensorineural hearing loss in the elderly. All other choices indicate a conductive hearing loss.

13. The answer is C. The superior salivatory nucleus contains the cell bodies of preganglionic parasympathetic neurons with axons in the facial nerve. There would be anterograde degeneration in the solitary nucleus; no other choice contributes fibers to the facial nerve.

14. The answer is D. The patient has a lesion of fibers in the sensory limb of the blink reflex on the left that are carried by the ophthalmic division of CN V. The ophthalmic division of CN V also supplies the forehead and scalp in addition to the cornea.

15. The answer is D. In warm-water caloric testing on the left, the eyes would drift slowly away from the irrigated side and the fast phase would be back to the irrigated side (warm/same).

16. The answer is C. All other structures are found in either the pons or midbrain.

17. F	31. D	45. C
18. O	32. H	46. E
19. L	33. E	47. F
20. C	34. K	48. A
21. Q	35. N	49. B
22. N	36. B	50. A
23. L	37. I	51. E
24. A	38. A	52. F
25. Q	39. U	53. C
26. P	40. B	54. D
27. N	41. A	55. B
28. L	42. J	56. C
29. G	43. B	57. E
30. S	44. B	

CHAPTER 5
CEREBELLUM AND BASAL GANGLIA

I. **The cerebellum and basal ganglia plan, initiate, and fine-tune movement generated by skeletal muscle contractions.**

  A. The **cerebellum functions in the planning and fine-tuning of movements**.
    1. The **cerebellum regulates movements** by comparing an intended performance with an actual performance.
    2. The **cerebellum attempts to reduce errors in the execution of movements**.

  B. The **basal ganglia function in the initiation of movement but also have cognitive and limbic functions**.

  C. Both the **cerebellum and basal ganglia mediate their effects by influencing upper motor neurons**.

  D. Both the **cerebellum and basal ganglia have excitatory inputs that use glutamate and inhibitory outputs that use** gamma-aminobutyric acid (GABA) as neurotransmitters.

  E. The **basal ganglia are influenced by dopaminergic and cholinergic neurons**, which do not play a role in the function of the cerebellum.

CEREBELLAR AND BASAL GANGLIA DISORDERS AND TREMOR

• *Patients with a lesion of the cerebellum or basal ganglia may have different forms of a tremor without paresis or paralysis of skeletal muscles.*
• *Lesions of cerebellar structures commonly result in a voluntary tremor evident with an intended movement.*
• *Lesions of basal ganglia structures commonly result in an involuntary tremor, or tremor at rest.*

  F. The **cerebellum influences upper motor neurons mainly in the contralateral cerebral cortex and brainstem**.

  G. The **basal ganglia influence upper motor neurons in the ipsilateral cerebral cortex**.

UNILATERAL LESIONS OF THE CEREBELLUM AND BASAL GANGLIA AND THE SIDE OF THE MOTOR DEFICITS

• *A **unilateral lesion of the cerebellum** may result in deficits in skeletal muscle performance ipsilateral to the side of the lesion (e.g., the right side of the cerebellum projects to upper motor neuron cell bodies on the left, but the axons of upper motor neurons [corticospinal] cross and influence lower motor neurons on the right). A patient may fall toward the side of the lesioned cerebellum.*

- *A **unilateral lesion of the basal ganglia** may result in deficits in skeletal muscle performance that are contralateral to the side of the lesion (e.g., the right basal ganglia project to upper motor neuron cell bodies on the right, but the axons of upper motor neurons [corticospinal] cross and influence lower motor neurons on the left).*

II. **The cortex of the cerebellum consists of a midline vermis, 2 lateral cerebellar hemispheres, and a flocculonodular lobe, which contain several topographic maps of the skeletal muscles in the body (Figure 5–1A and B).**

    A. **The midline vermis and the adjacent paravermal or intermediate regions of the cerebellar hemispheres** form part of the spinocerebellum that regulates ongoing execution of movements.

        1. The **vermis contains topographical maps of trunk muscles and proximal limb muscles**.

        2. The **intermediate parts of the hemisphere contain maps of distal limb muscles**.

        3. Most of the inputs to the vermis and intermediate hemisphere are from proprioceptive spinocerebellar pathways, which arise from skeletal muscles innervated by both spinal and cranial nerves.

    B. **The lateral part of each hemisphere forms part of the cerebrocerebellum that is mainly involved in the planning and rehearsal of skilled movements**, in particular those that involve distal limb movements.

        1. Most of the cerebellar inputs to the lateral hemisphere arise from the motor cortex and synapse in the pontine nuclei.

        2. The pontine nuclei send their axons across the midline of the pons and into the contralateral lateral hemisphere.

    C. The **flocculonodular lobe forms part of the vestibulocerebellum**.

        1. The flocculonodular lobe is involved in control of balance and the smooth execution of conjugate eye movements.

        2. Most of the input to the flocculonodular lobe arises from central processes of the vestibular nerve and from the vestibular nuclei.

    D. The **cerebellar cortex consists of 3 neuronal layers**, which contain 4 types of inhibitory neurons and 1 type of excitatory neuron (see Figure 5–1C).

        1. The **molecular layer** is the outer layer and contains basket cells and stellate cells.

            a. Both basket cells and stellate cells are inhibitory and use GABA as a neurotransmitter.

            b. Both basket cells and stellate cells inhibit Purkinje cells.

            c. The molecular layer contains parallel fibers, which are axons of granule cells (see later discussion).

        2. The **Purkinje cell layer** is the middle layer of the cerebellar cortex.

            a. Purkinje cells are inhibitory neurons that use GABA as a neurotransmitter and are the only neurons that send axons out of the cerebellar cortex.

            b. The dendrites of Purkinje cells extend into the molecular layer.

        3. The **granule cell layer** is the innermost layer of cerebellar cortex and contains Golgi cells and granule cells, which form synaptic glomeruli.

            a. Each glomerulus contains a cell body of a granule cell and axons of Golgi cells and mossy fibers that synapse with granule cells.

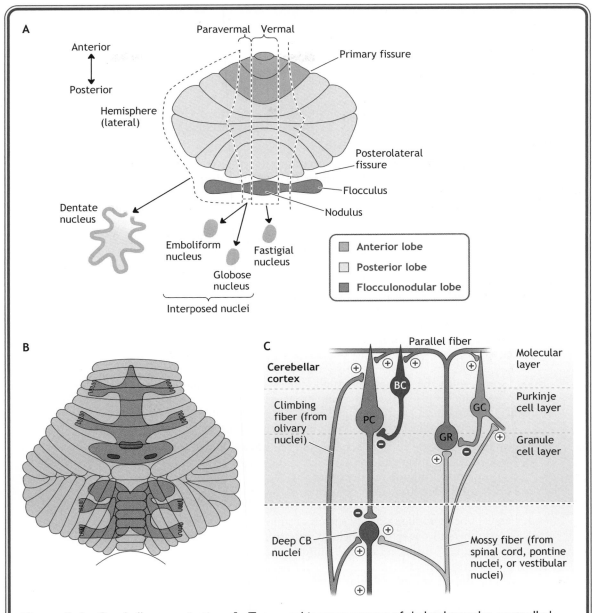

Figure 5–1. Cerebellar organization. **A:** Topographic arrangement of skeletal muscles controlled by parts of the cerebellum. **B:** Parts of the cerebellar cortex and the deep cerebellar nuclei linked together by Purkinje cells. **C:** Cytology of the cerebellar cortex.

b. Granule cells are the only excitatory neurons in the cerebellar cortex; their axons project into the molecular layer, bifurcate into parallel fibers, and convey mossy fiber input to Purkinje cell dendrites.

    c. Golgi cells are inhibitory neurons that are activated by parallel fiber axons of granule cells.

    d. Golgi cell axons provide recurrent inhibition back to granule cells in the glomeruli.

III. Inputs to the cerebellum directly or indirectly influence Purkinje cells (see Figure 5–1C).

  A. Climbing fibers provide a direct powerful monosynaptic excitatory input to Purkinje cells and also influence the activity of the deep cerebellar nuclei through axon collaterals.

    1. Climbing fibers encircle the cell bodies and proximal dendrites of Purkinje cells.

      a. Climbing fibers synapse with up to 10 Purkinje cells.

      b. Each Purkinje cell receives only a single climbing fiber.

    2. Climbing fibers are axons of neurons found in the inferior olivary complex of nuclei in the medulla.

    3. The **inferior olive** samples somatosensory, visual, auditory, and motor information and acts as a movement "error detector" for Purkinje cells.

    4. Climbing fibers generate prolonged calcium conductance in Purkinje cells, which results in the production of complex spike action potentials in Purkinje cell axons.

      a. Complex spikes generated in Purkinje cells may result in long-term depression in the strength of parallel fibers (axons of granule cells), which generate simple spikes at the same time in Purkinje cell axons.

      b. The long-term depression effects in the parallel fibers are the basis for motor or nondeclarative learning in the cerebellum.

  B. Mossy fibers provide an indirect diffuse excitatory input to Purkinje cells.

    1. Mossy fibers excite granule cells, which in turn give rise to parallel fiber axons that project to Purkinje cell dendrites.

    2. Purkinje cell dendrites may be contacted by as many as 1 million parallel fiber axons.

    3. Parallel fibers produce a steady stream of simple spikes in Purkinje cell axons.

    4. Mossy fibers to the spinocerebellum arise from muscle stretch receptors and Golgi tendon organs in skeletal muscles and project to the cerebellum in spinocerebellar tracts.

      a. The **dorsal spinocerebellar tract** arises from Clarke's nucleus from the T1 through L2 spinal cord segments and relays information about the biomechanical state of muscle spindles and Golgi tendon organs in skeletal muscles in the lower limb.

      b. The **cuneocerebellar tract** arises from neurons in the external cuneate nucleus in the medulla and relays information about muscle spindles and Golgi tendon organs in skeletal muscles in the upper limb.

      c. The **ventral spinocerebellar tract** (VSCT) arises from spinal border cells adjacent to lamina VII of the spinal cord and conveys information to the cerebellum about skeletal muscle activity generated by upper motor neurons.

5. Mossy fibers to the cerebrocerebellum arise from pontine nuclei, which are influenced by axons of neurons in the cerebral cortex.
6. Mossy fibers to the vestibulocerebellum arise from the vestibular nerve and the vestibular nuclei.

IV. **The cerebellum is connected to the brainstem by the inferior, middle, and superior cerebellar peduncles, which convey axons into and out of the cerebellum (Figures 4–1B, 5–2, and 5–3).**

A. The **superior cerebellar peduncle mainly conveys axons out of the cerebellum from the deep cerebellar nuclei.**

B. The **middle cerebellar peduncle conveys axons from the pontine nuclei into the cerebellum**.

C. The **inferior cerebellar peduncle conveys axons both into the cerebellum from the spinal cord and brainstem and out of the cerebellum from the deep cerebellar nuclei.**

V. **The output of the cerebellar cortex is provided exclusively by axons of Purkinje cells (see Figures 5–1A and 5–2).**

A. **Purkinje cell axons leave the cerebellar cortex and project to the deep cerebellar nuclei or the vestibular nuclei** in a topographically ordered fashion.

B. The **deep cerebellar nuclei are, medial to lateral, the fastigial nucleus, the interposed nuclei (globose and emboliform), and the dentate nucleus** (see Figures 5–1A, 5–2, and 5–3).
   1. Purkinje cell axons in the spinocerebellum (vermis and intermediate hemisphere) influence the activity of the fastigial and interposed nuclei.
      a. Purkinje cell axons in the vermis project to the fastigial nuclei.
      b. Purkinje cell axons in the intermediate hemisphere project to the interposed (globose and emboliform) nuclei.
   2. Purkinje cell axons in the cerebrocerebellum (the lateral part of the hemisphere) project to the dentate nucleus (see Figure 5–2).
   3. Purkinje cell axons in the vestibulocerebellum (flocculonodular lobe) project to the fastigial nucleus and to the lateral and medial vestibular nuclei.

C. **Axons of deep cerebellar nuclei exit the cerebellum mainly in the superior cerebellar peduncle and influence upper motor neurons in the cerebral cortex and brainstem** (see Figure 5–2).
   1. Axons from the dentate, fastigial, and interposed nuclei exit in the superior cerebellar peduncle, cross the midline in the midbrain, and terminate in the ventral lateral nucleus of the thalamus.
   2. The thalamus projects to the motor cortex and influences corticospinal and corticobulbar upper motor neurons.
   3. Axons from the interposed nuclei also project to the red nucleus and influence upper motor neurons in the rubrospinal tract.
   4. Axons from the fastigial nuclei influence reticulospinal and vestibulospinal upper motor neurons in the reticular formation and vestibular nuclei.

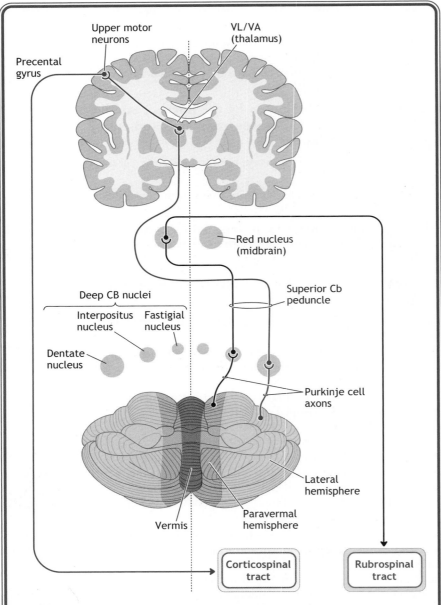

Figure 5–2. Cerebrocerebellar circuitry illustrating how Purkinje cells in the paravermal and lateral hemispheres are linked to upper motor neurons.

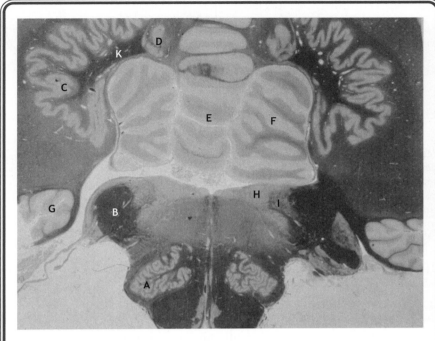

Figure 5–3. Myelin-stained section of the medulla and overlying cerebellar structures. **A:** Inferior olive. **B:** Inferior cerebellar peduncle. **C:** Dentate nucleus. **D:** Interposed nuclei. **E:** Vermis. **F:** Tonsil of paravermal hemisphere. **G:** Flocculus. **H:** Medial vestibular nucleus. **I:** Lateral vestibular nucleus.

LESIONS OF THE VERMIS AND SPINOCEREBELLUM

- Patients with **lesions of the vermal region** may have difficulty maintaining posture or balance and have a motor or cerebellar ataxic gait.
- Patients with **motor ataxia** can be differentiated from those with sensory ataxia (which may result from a lesion of the dorsal columns) by the absence of a Romberg sign.
- In **cerebellar lesions**, patients will sway or lose their balance with their eyes open or closed. In dorsal column lesions that result in sensory ataxia, **patients sway with their eyes closed, a positive Romberg sign**.

LESIONS OF THE FLOCCULONODULAR LOBE AND THE VESTIBULOCEREBELLUM

- Patients with **lesions of the flocculonodular lobe** may have **scanning dysarthria**, which causes patients to divide words into individual syllables, disrupting the melody of speech.
- Patients may have abnormal conjugate gaze. When the eyes move and try to fix on a target, they may pass it or stop too soon and oscillate a few times before they settle on the target. Patients may have coarse cerebellar nystagmus, with a fast phase, which is usually directed toward the side of the lesion.

NEURONS IN THE VERMIS AND THIAMINE DEFICIENCY

- **Purkinje cells in the anterior part of the vermis** that control proximal muscles in the lower limbs are preferentially affected in patients with a thiamine deficiency associated with malnourishment or alcoholism.
- These patients have pronounced motor ataxia but no cerebellar signs in the upper limbs or cranial musculature.

LESIONS OF A CEREBELLAR HEMISPHERE AND A TREMOR WITH MOVEMENT

- In patients with **hemisphere lesions**, an intention tremor is seen when voluntary movements are performed and increases as a target is approached; the tremor is barely noticeable or absent at rest.
- These patients may have **dysmetria**, an inability to stop a movement at the proper place; the patient has difficulty performing a finger-to-nose test.
- These patients may have **dysdiadochokinesis**, a decreased ability to perform rapidly alternating movements, such as pronation and supination of the forearm with slapping of the anterior thigh between movements.

VI. **The basal ganglia participate in 4 "direct" pathways: motor, oculomotor, cognitive, and limbic (Figure 5–4).**

  A. **The direct basal ganglia pathways are driven by excitatory inputs from glutamatergic neurons in large areas of the cerebral cortex.**

  B. **The direct basal ganglia pathways influence thalamic nuclei, which project back to specific areas of the frontal lobe.**

  C. **The direct basal ganglia pathways function in parallel with an indirect basal ganglia pathway.**

  D. Both **direct and indirect basal ganglia pathways contain 2 populations of GABA neurons**, which interact to create a disinhibition of neurons in the diencephalon and cortex.

  E. Both **direct and indirect basal ganglia pathways are influenced by dopaminergic neurons and cholinergic neurons.**

VII. **The components of the basal ganglia consist of several subcortical telencephalic structures, including the striatum and globus pallidus, and a midbrain nucleus, the substantia nigra (Figures 5–4 and 5–5).**

  A. The **striatum** consists of a large dorsal component, which includes the caudate nucleus and the putamen, and a small ventral part, which includes the nucleus accumbens.

  1. The **caudate nucleus** has a C-shaped configuration and consists of a head, body, and tail; the caudate nucleus is separated from the putamen by the anterior limb of the internal capsule.

  2. The **caudate nucleus and putamen** are histologically similar and contain medium-sized spiny GABA neurons.

  3. The **ventral striatum**, which contains the nucleus accumbens, is situated at the point where the head of the caudate is continuous with the putamen ventral to the internal capsule.

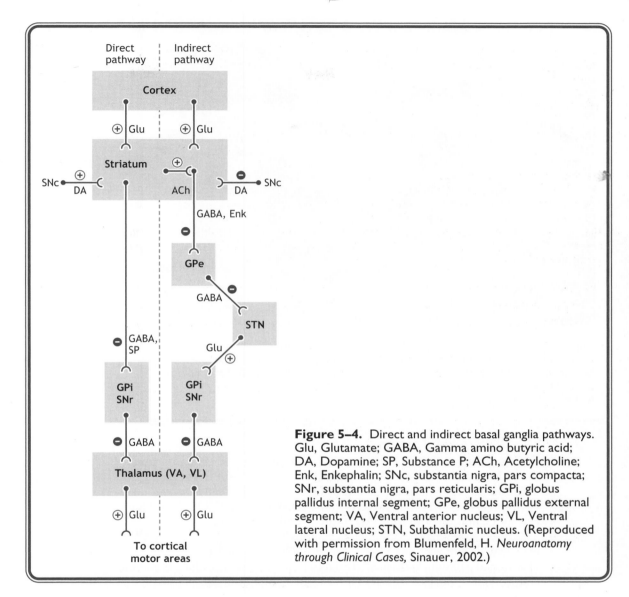

Figure 5–4. Direct and indirect basal ganglia pathways. Glu, Glutamate; GABA, Gamma amino butyric acid; DA, Dopamine; SP, Substance P; ACh, Acetylcholine; Enk, Enkephalin; SNc, substantia nigra, pars compacta; SNr, substantia nigra, pars reticularis; GPi, globus pallidus internal segment; GPe, globus pallidus external segment; VA, Ventral anterior nucleus; VL, Ventral lateral nucleus; STN, Subthalamic nucleus. (Reproduced with permission from Blumenfeld, H. *Neuroanatomy through Clinical Cases,* Sinauer, 2002.)

    4. The striatum contains cholinergic neurons, which inhibit GABA neurons in the direct pathway and excite GABA neurons in the indirect pathway.

B. The **external and internal segments of the globus pallidus and the ventral pallidum** receive most of their inputs from the striatum.

    1. The **external and internal segments of the globus pallidus are situated medial to the putamen**. Together, these 3 structures form the lentiform nuclei.

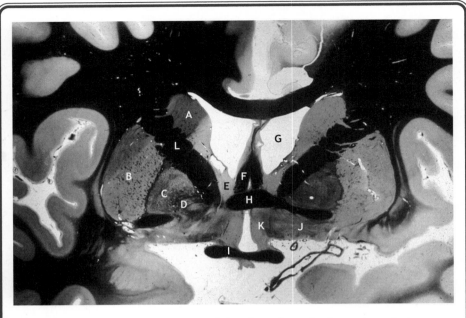

Figure 5–5. Coronal section through basal ganglia and other subcortical structures. **A:** Caudate nucleus. **B:** Putamen. **C:** Globus pallidus external segment. **D:** Globus pallidus internal segment. **E:** Septal nuclei. **F:** Fornix. **G:** Lateral ventricle. **H:** Anterior commissure. **I:** Optic chiasm. **J:** Basal nucleus of Meynert. **K:** Preoptic hypothalamus. **L:** Internal capsule, anterior limb.

   2. Both segments of the globus pallidus and the ventral pallidum contain GABA neurons.

   3. The internal segment of the globus pallidus is a component of several direct basal ganglia pathways and projects to the **ventral lateral (VL), ventral anterior (VA), and mediodorsal (MD) nuclei** of the thalamus.

   4. The external segment of the globus pallidus is a component of an indirect basal ganglia pathway and projects to the subthalamic nucleus.

   5. The ventral pallidum is situated ventral to the segments of the globus pallidus and contains the basal nucleus of Meynert, a prominent population of cholinergic neurons.

C. The **substantia nigra in the midbrain consists of a pars reticularis and a pars compacta** (see Figure 5–4).

   1. The substantia nigra, pars reticularis, contains GABA neurons and is histologically similar to the globus pallidus internal segment.

   2. The substantia nigra, pars compacta, and an adjacent ventral tegmental area contain dopamine neurons, which project into the direct and indirect basal ganglia pathways.

      a. Dopamine binds with D_1 receptors, which enhance the direct basal ganglia pathways.

b. Dopamine binds with D_2 receptors, which suppress the indirect ganglia pathway.
   D. Four diencephalic nuclei contribute to basal ganglia pathways (see Figure 5–4).
   1. The VL, VA, and MD nuclei of the dorsal thalamus are excited by direct basal ganglia pathways and project to specific areas of the frontal lobe.
   2. The subthalamic nucleus contains glutamate neurons, which project back to the globus pallidus internal segment in the indirect basal ganglia pathway.

VIII. The basal ganglia direct motor pathway drives or excites motor cortex and promotes the initiation of movement (see Figure 5–4).

   A. In the direct motor pathway, excitatory input from primary motor and premotor cortex projects to GABA neurons, mainly in the putamen, which also contain the neuropeptide substance P.
   1. Activated GABA neurons in the striatum inhibit GABA neurons in the internal segment globus pallidus and substantia nigra, pars reticularis; this disinhibits and excites the VA and VL nuclei of the thalamus.
   2. The VA and VL thalamic nuclei excite upper motor neurons in all areas of motor cortex and promote the initiation of movement.
   3. Axons of dopamine neurons in the substantia nigra, pars compacta, form a nigrostriatal or mesostriatal pathway, which excites the direct pathway by binding with D_1 receptors on striatal GABA-substance P neurons.
   B. The basal ganglia motor pathway is balanced by the indirect basal ganglia pathway (Figure 5–4).
   1. In the indirect pathway, excitatory input from the cerebral cortex projects to GABA neurons, which also contain the neuropeptide enkephalin in the caudate nucleus and putamen.
   2. The GABA neurons in the striatum inhibit a second population of GABA neurons in the external segment of the globus pallidus, which disinhibits and excites the subthalamic nucleus.
   3. The glutamatergic neurons in the subthalamic nucleus excite GABA neurons in the internal segment of the globus pallidus.
   4. The internal segment inhibits the VA and VL thalamic nuclei, which reduces activity in motor cortex and suppresses the direct basal ganglia pathways.
   5. The effects of the indirect basal ganglia pathway may be enhanced by cholinergic neurons in the caudate nucleus and putamen.
   6. Axons of dopamine neurons inhibit the indirect pathway by binding with D_2 receptors on striatal GABA-enkephalin neurons.

LESIONS OR DISEASES OF THE DIRECT MOTOR BASAL GANGLIA PATHWAYS

• *Patients with **lesions or diseases of the direct motor basal ganglia pathways** have movement disorders known as dyskinesias and an involuntary tremor, or tremor at rest.*
• *Lesions of the direct pathway result in an underactive motor cortex and hypokinetic disturbances.*
  *-**The most common signs and symptoms of a direct pathway lesion result from Parkinson's disease**, which is caused by the degeneration of dopaminergic neurons in the substantia nigra, pars compacta.*
  -The motor cortex is suppressed in patients with Parkinson's disease; patients have problems initiating movements, combined with reduced velocity and amplitude of the movements.

-A **"pill-rolling" tremor** at rest is seen in the fingers.

-Skeletal muscles in the upper limbs may exhibit a **cogwheel**, or "lead pipe," **rigidity** because of increased muscle tone. Recall that patients with upper motor neuron lesions have increased muscle tone described as spasticity.

-Patients typically have a stooped posture, an expressionless face, and a shuffling or accelerating gait during which individuals seem to chase their center of gravity.

• **Lesions or diseases of the indirect pathway** (chorea, athetosis, dystonia, tics, hemiballismus) result in an overactive motor cortex, hyperkinetic disturbances, and pronounced involuntary movements.

-**Chorea** refers to involuntary movements that are purposeless, quick jerks that may be superimposed on voluntary movements. **Huntington's chorea** exhibits autosomal dominant inheritance (chromosome 4) and is characterized by degeneration of indirect pathway GABA neurons and AEh neurons in the striatum, in particular the head of the caudate nucleus. **Sydenham's chorea** is a transient complication observed in some children with rheumatic fever.

-**Athetosis** refers to slow, writhing, involuntary movements that are most evident in the fingers and hands. It is frequently seen in patients with Huntington's disease and may be observed in any disease that involves the indirect pathway.

-**Dystonia** refers to slow, prolonged movements mainly involving the trunk musculature, and it often occurs with athetosis. Examples of dystonic movements include **blepharospasms** (contractions of the orbicularis oculi muscles, causing the eyes to close), **spasmodic torticollis** (in which the head is pulled toward the shoulder), and **writer's cramp** (contraction of arm and hand muscles on attempting to write).

-**Hemiballismus** results from a lesion of a subthalamic nucleus. It is commonly caused by a lacunar stroke of a thalamoperforating branch of a posterior cerebral artery in hypertensive patients. Patients exhibit violent projectile ballistic movements of a limb. Hemiballismus is observed in the limbs contralateral to the lesioned subthalamic nucleus.

-Patients with **Tourette's syndrome** have facial and vocal tics that progress to jerking movements of the limbs. They may also have involuntary explosive, vulgar speech.

-**Wilson's disease** is caused by an abnormality of copper metabolism. It results in the accumulation of copper in the liver. Patients may have **personality changes, dystonia, and a "wing-beating" tremor**. A thin, brown **Kayser-Fleischer ring** may be present around the outer cornea, aiding in the diagnosis. Lacunar degeneration in the putamen is pathoanatomic.

IX. The oculomotor direct basal ganglia pathway functions in the initiation of horizontal and vertical saccadic eye movements.

A. In the oculomotor pathway, **excitatory input from the posterior parietal (visual association) cortex projects to GABA neurons,** mainly in the body of the caudate nucleus.

B. Activated GABA neurons in the caudate inhibit GABA neurons in the substantia nigra, pars reticularis; this disinhibits and excites the VA and MD nuclei of thalamus.

C. GABA neurons in the substantia nigra, pars reticularis, project to the superior colliculus; disinhibition of the superior colliculus directly promotes the initiation of saccades.

D. The VA and MD thalamic nuclei also excite neurons in the frontal eye fields, which promote the initiation of saccades.

X. The cognitive direct basal ganglia pathway functions in "executive" behaviors, including organizing verbal skills in problem solving and coordinating socially appropriate responses.

A. In the cognitive, or executive, pathway, excitatory input from the prefrontal cortex projects to GABA neurons mainly in the head of the caudate nucleus.

B. Activated GABA neurons in the striatum inhibit GABA neurons in the internal segment globus pallidus and substantia nigra, pars reticularis; this disinhibits and excites the VA and MD nuclei of the thalamus.

C. The VA and MD thalamic nuclei project back to the dorsolateral prefrontal cortex.

D. Axons of dopamine neurons in the ventral tegmental area form a mesocortical projection, which drives the cognitive pathway.

XI. The limbic direct basal ganglia pathway functions in motivational behavior and in enhancing emotional memories.

A. The **limbic pathway is driven by neurons in the temporal lobe**, including the entorhinal cortex, hippocampus, and amygdala, which project to the nucleus accumbens in the ventral striatum.

B. The **ventral striatum projects to the ventral pallidum** (ventral part of the globus pallidus internal segment and the basal nucleus of Meynert).

C. **Activated neurons in the ventral pallidum excite the MD nucleus of the thalamus.**

D. The **MD thalamic nucleus projects to limbic cortical areas in the anterior cingulate gyrus and orbitofrontal region**.

E. Axons of dopamine neurons in the ventral tegmental area form a **mesolimbic pathway**, which drives the limbic pathway.

HUNTINGTON'S DISEASE

- **Huntington's disease** may affect the cognitive and limbic basal ganglia pathways.
- Patients with Huntington's disease may exhibit **changes in mood or character** in the form of irritability or impulsive behavior because of a loss of GABA neurons in the cognitive pathway.

SCHIZOPHRENIA

- **Schizophrenia** may result from changes in the activity of the mesolimbic and mesocortical dopamine projections in the cognitive and limbic pathways.
  -Schizophrenia is characterized by **positive and negative psychotic symptoms**. Positive symptoms include delusions, disordered thoughts with incoherence, loss of touch with reality, and hallucinations. The hallucinations may be auditory, in which patients report hearing voices making untrue statements. The negative symptoms of schizophrenia include social withdrawal, lack of motivation, poor attention span, slowed speech, and a lack of emotional arousal.
  -An overactive mesolimbic dopaminergic pathway may cause the positive symptoms of schizophrenia. Most antipsychotic drugs have an affinity with D_2 dopamine receptors, which decrease dopamine transmission and reduce the positive symptoms.
  -Decreased activity in the mesocortical dopaminergic pathway may cause the negative symptoms of schizophrenia.
  -Patients undergoing treatment with antipsychotic drugs that alter or reduce dopamine transmission in the motor basal ganglia pathway may develop tardive dyskinesia. They have involuntary movements of the tongue and face and cogwheel rigidity similar to that seen in Parkinson's disease patients.

OBSESSIVE-COMPULSIVE DISORDER

- **Obsessive-compulsive disorder** may result from overactivity in the striatal part of the cognitive basal ganglia pathway.
  -Patients have recurrent obsessions and attempt to overcome the anxiety associated with the obsessions by performing compulsive acts. Obsessive-compulsive disorder may be caused by hyperactivity of GABA neurons in the head of the caudate nucleus.

DRUGS OF ABUSE AND ADDICTION

- **Drugs of abuse and addiction** enhance the effects of dopamine in the mesolimbic projection to the nucleus accumbens. Cocaine, amphetamines, and morphine increase the time that dopamine remains in the synaptic cleft. Nicotine acts by enhancing the release of dopamine.

CLINICAL PROBLEMS

1. Both basal ganglia pathways use disinhibition to mediate their effects. Which of the following pairs of structures contain the 2 inhibitory neuron cell bodies that interact to create a disinhibition?

    A. Substantia nigra, pars compacta—putamen

    B. Globus pallidus external segment—subthalamic nucleus

    C. Globus pallidus internal segment—VA nucleus of the thalamus

    D. VA nucleus of the thalamus—motor cortex

    E. Putamen—globus pallidus internal segment

2. A 55-year-old male patient develops dancelike involuntary movements. You diagnose the patient as having a degenerative neurological disease that was also evident in the patient's father and uncle. Where is the most likely site of the degeneration?

    A. Globus pallidus, internal segment

    B. VL nucleus of the thalamus

    C. Head of caudate nucleus

    D. Substantia nigra, pars compacta

    E. Putamen

3. A hypertensive patient suffers a lacunar infarct and develops uncontrollable violent flinging movements of the left upper limb. Where is the lesion?

    A. Left globus pallidus internal segment

    B. Right subthalamic nucleus

    C. Left substantia nigra

    D. Right VL nucleus

    E. Right substantia nigra

4. Over a period of years, a construction worker begins to have difficulty hammering nails using his right arm and hand. When he walks, he seems to teeter to the right. A

neurological exam reveals that muscle strength is 5/5 in both the upper and lower limbs, but the patient exhibits a tremor when attempting to touch his right index finger to the tip of his nose. Cranial nerve testing and the rest of the neurological exam are normal. Where might a lesion be located that accounts for all of the patient's symptoms?

   A. Anterior part of the vermis bilaterally

   B. Lateral part of the hemisphere on the right

   C. The vermis and cerebellar hemisphere on the right

   D. Head of the caudate nucleus on the right

   E. Inferior olivary nuclei on the right

MATCHING PROBLEMS

Questions 5–13: Motor control systems match: Use the following structures to answer the questions that follow.

Choices (each choice may be used once, more than once, or not at all):

   A. Vermis of cerebellum

   B. Head of caudate nucleus

   C. Globus pallidus external segment

   D. Globus pallidus internal segment

   E. Dentate nucleus

   F. Subthalamic nucleus

   G. Substantia nigra

   H. More than one choice in A–G is an appropriate answer

  5. Contains cell bodies of neurons that utilize GABA as a neurotransmitter.

  6. Axons of cell bodies here synapse in the thalamus.

  7. Axons from neuron cell bodies here cross the midline.

  8. Contains cells bodies of neurons that utilize dopamine as a neurotransmitter.

  9. A neuronal structure utilized by both the direct and indirect basal ganglia pathways.

 10. Contains granule cells.

 11. Contains cell bodies of neurons that utilize acetylcholine as a neurotransmitter.

 12. Climbing fibers synapse here.

 13. A derivative of the diencephalon.

Questions 14–29: Basal ganglia match.

Choices (each choice may be used once, more than once, or not at all):

   A. Direct basal ganglia pathway

   B. Indirect basal ganglia pathway

C. Neither basal ganglia pathway

    D. Both basal ganglia pathways

14. Utilizes disinhibition of the subthalamic nucleus to mediate its effects.

15. Utilizes neurons in the external segment of the globus pallidus.

16. Effects are enhanced by dopamine.

17. Utilizes neurons in the diencephalon.

18. Excited by axons of glutamate neurons from cerebral cortex.

19. Lesions here result in an underactive motor cortex.

20. Effects are enhanced by acetylcholine.

21. Utilizes GABA neurons in its pathway.

22. Lesions here result in an overactive motor cortex.

23. Utilizes neurons in the internal segment of the globus pallidus.

24. Lesions here result in difficulty initiating movements.

25. A lesion here results in ballistic-like movements.

26. A lesion here results in chorea.

27. Utilizes disinhibition of the VL nucleus of the thalamus to mediate its effects.

28. A unilateral lesion here results in effects that are ipsilateral to the lesion.

29. A lesion here results in a shuffling gait and a pill-rolling tremor.

ANSWERS

1. The answer is E. The putamen–globus pallidus internal segment contains GABA neurons that interact to create a disinhibition in the direct basal ganglia pathway and excitation of the thalamus and cortex.

2. The answer is C. The patient has Huntington's disease. Huntington's chorea exhibits autosomal dominant inheritance (chromosome 4) and is characterized by degeneration of indirect pathway GABA neurons in the striatum, in particular the head of the caudate nucleus.

3. The answer is B. The patient has hemiballismus, which results from a lesion of a subthalamic nucleus and is commonly caused by lacunar stroke of a thalamoperforating branch of a posterior cerebral artery in hypertensive patients. Patients with hemiballismus exhibit violent projectile ballistic movements of a limb which are observed in the limbs contralateral to the lesioned subthalamic nucleus, in this case the right.

4. The answer is C. This patient has an intention tremor with dysmetria, implicating the right hemisphere, and an ataxic gait that involves the vermis on the same side of the intention tremor.

5. H (A–D apply)

6. H (D and E apply)

7. E

8. G

9. D

10. A

11. B

12. H (A and E apply)

13. F

14. B

15. B

16. A

17. D

18. D

19. A

20. B

21. D

22. B

23. D

24. A

25. B

26. B

27. A

28. C

29. A

CHAPTER 6
DIENCEPHALON AND LIMBIC SYSTEM

I. **The diencephalon consists of the (dorsal) thalamus, hypothalamus, epithalamus, and subthalamus.**

II. **The thalamus is mainly a collection of nuclei to which virtually all sensory and motor systems project on their way to the cerebral cortex (Figure 6–1A and B).**

A. Virtually all **thalamic nuclei project to some part of the cerebral cortex,** and that region of the cortex projects back to the same thalamic nuclei.
   1. The dorsal column—the medial lemniscal system, the anterolateral system, and the trigeminal system convey somatosensory information to the cortex through nuclei of the thalamus.
   2. Motor output of the basal ganglia and cerebellum is relayed to the cerebral cortex through motor nuclei of the thalamus.

B. Each **thalamus is an egg-shaped structure that is separated by the third ventricle**; the 2 thalami may be joined across the midline by an interthalamic adhesion.

C. Within the thalamus, the **internal medullary lamina divides the thalamus into anterior, medial, and lateral regions** (see Figure 6–1A and B).
   1. The anterior region is situated between the fibers of the internal medullary lamina and contains the anterior nucleus.
   2. The medial region is medial to the internal medullary lamina and contains the mediodorsal nucleus.
   3. The lateral region can be divided into a ventral tier and a dorsal tier of nuclei.
      a. The dorsal tier contains the lateral dorsal (LD) and lateral posterior (LP) nuclei and the pulvinar.
      b. The ventral tier contains the ventral anterior and ventral lateral nuclei and the ventrobasal complex, consisting of the ventral posterior lateral (VPL) and ventral posterior medial (VPM) nuclei.
      c. The lateral geniculate nucleus (LGN) and medial geniculate nucleus (MGN) are suspended from the ventral tier near the caudal pole of the thalamus.

D. **Thalamic nuclei may be classified as either specific or nonspecific or as a relay or a generalized nucleus**; neither scheme deals effectively with all thalamic nuclei.

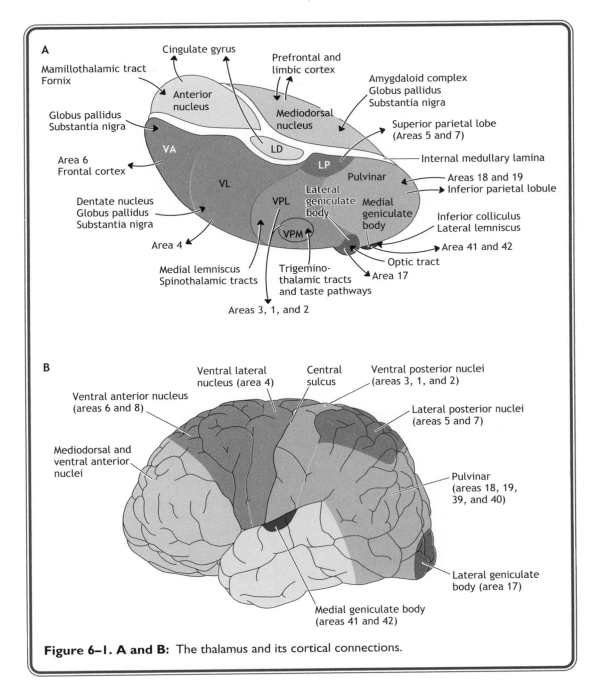

Figure 6–1. A and B: The thalamus and its cortical connections.

1. Specific thalamic nuclei receive their input from a specific sensory or motor system; nonspecific nuclei receive inputs from several sources.
2. Relay thalamic nuclei project to restricted, functionally specific regions of the cerebral cortex; generalized nuclei project to multiple regions of association cortex.
3. Specific nuclei are usually relay nuclei, which are designed for preservation of information transfer from specific motor or sensory systems to the cortex.
4. Nonspecific nuclei are usually generalized nuclei, which are involved in parlaying 1 or more types of input to broad areas of the cerebral cortex.

E. Specific or relay nuclei (see Figure 6–1A and B) include the LGN, MGN, ventrobasal complex, anterior nucleus, and ventral lateral nucleus.
1. The LGN is a 6-layered nucleus that receives bilateral topographically organized input from the retina of each eye.
   a. Retinal ganglion cells send their axons to the LGN via the optic nerve, chiasm, and tract.
   b. At the chiasm, a partial crossing of optic nerve fibers occurs so that each geniculate is receiving input from both retinas.
   c. The layers maintain the segregated input from the 2 eyes; the contralateral retina projects to layers 1, 4, and 6, and the ipsilateral retina projects to layers 2, 3, and 5.
   d. The dorsal (parvocellular) layers (3–6) of the LGN receive input that is processing form and color, and the deeper (magnocellular) layers (1–2) receive input related to motion, depth, and spatial information.
   e. The LGN projects in a topographically ordered manner to the primary visual cortex in the occipital lobe.
2. The MGN receives bilateral tonotopically organized auditory input from the inferior colliculus and projects to the primary auditory cortex, which is located in the superior temporal gyrus of the temporal lobe.
3. The ventrobasal complex includes the VPL and the VPM. It receives somatotopically organized input from the somatosensory systems from the limbs, neck, trunk, and the head and projects to the primary somatosensory cortex (see Figure 6–1A and B).
   a. The VPL receives axons that arise from neurons in the dorsal column nuclei (cuneatus and gracilis), which convey discriminative touch, vibration, pressure, and conscious proprioceptive information from the contralateral limbs and body by way of the medial lemniscus.
   b. The VPL receives axons from the spinothalamic tract (part of the anterolateral system), which conveys pain and temperature information from the contralateral limbs and body.
   c. Each medial lemniscal or spinothalamic neuron projects to a modality-specific neuron in the VPL; some VPL neurons are multimodal, receiving input from both sensory systems.
   d. The VPM is the site of termination of touch, proprioceptive, and pain and temperature information from the face, scalp, oral and nasal cavities, and dura. The VPM also relays taste information from the solitary nucleus to somatosensory cortex.
   e. The main sensory nucleus of V sends touch and proprioceptive information to the VPM by way of the dorsal trigeminothalamic tract.

f. The spinal trigeminal nucleus relays pain and temperature information by way of the ventral trigeminothalamic tract.

  g. The dorsal trigeminothalamic tract projects bilaterally to each VPM from the main sensory nuclei, whereas the ventral trigeminothalamic tract arises exclusively from the contralateral spinal nucleus.

  h. The ventral posterior inferior nucleus is a small group of neurons situated between the VPL and the VPM that receive input from the vestibular nuclei and projects to the postcentral gyrus in the depths of the central sulcus.

 4. The anterior nucleus receives afferents primarily from the mammillary nucleus in the hypothalamus and projects to the cingulate cortex; this projection forms part of the Papez circuit of limbic connections (see Figure 6–1A and B).

 5. The VL receives projections from the deep cerebellar nuclei and from the internal segment of the globus pallidus and substantia nigra, pars reticularis, of the basal ganglia.

  a. The cerebellar part of VL projects to the neurons in the precentral gyrus (primary motor cortex).

  b. The basal ganglia part of VL projects to broader areas of the frontal lobe, including supplementary motor and premotor areas.

  c. Cerebellar inputs to the VL arise from the contralateral deep cerebellar nuclei but basal ganglia inputs arise from the ipsilateral globus pallidus and substantia nigra.

 6. The ventral anterior nucleus receives virtually all of its input from the substantia nigra, pars reticularis, and the internal segment of the globus pallidus.

F. Nonspecific or generalized nuclei (see Figure 6–1A) include the **mediodorsal nucleus**, **lateral dorsal nucleus**, and the **lateral posterior nucleus**.

 1. The mediodorsal nucleus receives inputs from the amygdala, a part of the limbic system, and the basal ganglia and projects to most of the frontal lobe.

 2. The LD nucleus is similar to the anterior nucleus.

  a. The LD receives input primarily from the hypothalamus and projects to the cingulate cortex.

  b. The LD contributes an autonomic component to emotional processing in the limbic system.

 3. The LP nucleus and pulvinar act as visual association nuclei.

  a. These nuclei receive inputs from the superior colliculus, pretectal nuclei, and visual cortex.

  b. They project to visual association areas in the occipital, parietal, and temporal lobes.

G. Intralaminar nuclei are situated within the internal medullary lamina and include the midline nuclei and the centromedian nucleus.

 1. The midline nuclei receive inputs from the reticular formation and pain and temperature input from the spinoreticular component of the anterolateral system.

 2. The centromedian nucleus is the largest of the intralaminar nuclei. It receives inputs from the motor cortex and the basal ganglia and projects back to the basal ganglia.

H. The **reticular nucleus** is a sheet of cells on the outside of the external medullary lamina, where it encapsulates most of the rostral, ventral, and lateral surfaces of the dorsal thalamus.

1. The reticular nucleus filters information passing from the thalamus to the cortex and from the cortex back to the thalamus.
2. The reticular nucleus receives excitatory collaterals from the thalamocortical and corticothalamic axons, which project into and out of each dorsal thalamic nucleus.
3. The reticular nucleus contains gamma-aminobutyric acid (GABA) neurons, which provide an inhibitory topographic feedback to nuclei of the thalamus.

THALAMIC PAIN SYNDROME

- *Thalamic pain syndrome* commonly results from a lacunar stroke involving thalamoperforating branches of a posterior cerebral artery, which supply the ventrobasal complex.
- In these patients, there may be an **impairment of all forms of somatic sensations in the body and limbs** contralateral to the affected thalamus.
- Central pain may also occur if the lesion affects the anterolateral system part of the ventrobasal complex. In these patients, an initial analgesia is replaced over time by spontaneous aching, burning pain, which is often excruciating, and a heightened sensation of pain is felt, even after light cutaneous stimulation of the contralateral limbs or body. Frequently, there is a heightened sensation of cold or heat. Pain in thalamic pain syndrome will not respond to anti-inflammatory analgesics, which act to suppress pain at brainstem or spinal cord levels.

III. **The hypothalamus maintains homeostasis by controlling endocrine and autonomic functions and by interacting with the limbic system (mnemonic: HEAL).**

A. The **hypothalamus** can be organized in the sagittal plane into **anterior, middle, and posterior regions** (Figure 6–2A and B).

B. The anterior region is situated above or rostral to the optic chiasm and contains 4 main nuclei: preoptic, paraventricular, supraoptic, and suprachiasmatic.

1. The preoptic nuclei are actually telencephalic structures that are situated rostral to the optic chiasm adjacent to the lamina terminalis, the rostral boundary of the diencephalon.

   a. The medial preoptic nucleus produces **gonadotropin-releasing hormone** (GnRH). GnRH is released into the tuberoinfundibular tract and is transported through the hypophyseal portal system to the adenohypophysis.

   b. The medial preoptic nucleus is a sexually dimorphic nucleus. It is large in males (its neurons continually release GnRH) and small in females (its neurons release GnRH on a cyclical basis).

   c. The ventral lateral preoptic nucleus of the hypothalamus (VLPO) contains neurons that inhibit histaminergic neurons in the hypothalamus and cholinergic and noradrenergic neurons in the brainstem, helping to initiate the onset of non-rapid eye movement (non-REM) sleep.

2. The paraventricular and supraoptic nuclei produce and secrete the neuropeptides vasopressin (ADH) and oxytocin. Axons arising from these nuclei leave the hypothalamus and carry these neurosecretory products to the neurohypophysis.

   a. The release of vasopressin is regulated by chemoreceptors in the carotid body and by the subfornical organ and the organum vasculosum, 2 circumventricular organs situated in the walls of the third ventricle.

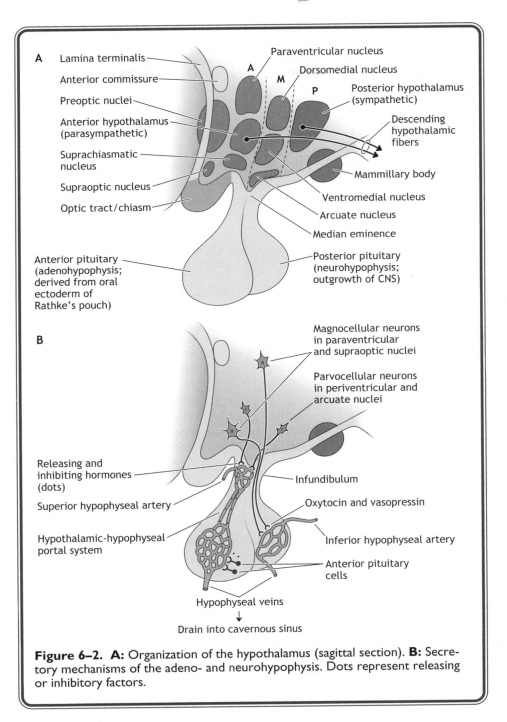

Figure 6–2. A: Organization of the hypothalamus (sagittal section). **B:** Secretory mechanisms of the adeno- and neurohypophysis. Dots represent releasing or inhibitory factors.

b. The subfornical organ is sensitive to circulating levels of angiotensin II, which is derived from the production of renin by the kidneys. Angiotensin II causes an increase in the release of vasopressin and helps maintain blood volume and blood pressure.
   c. The organum vasculosum responds to low plasma osmolarity by causing an increase in the release of vasopressin.
   d. The paraventricular nucleus also secretes **corticotropin-releasing hormone** (CRT).

LESIONS OF THE SUPRAOPTIC AND PARAVENTRICULAR NUCLEI OR TRAUMA TO THE HYPOTHALAMOHYPOPHYSIAL TRACT

*Lesions of the supraoptic and paraventricular nuclei or trauma to the hypothalamohypophysial tract may result in **diabetes insipidus**, which is characterized by polydipsia (excess water consumption) and polyuria (excess urination).*

   3. The suprachiasmatic nucleus receives direct input from retinal ganglion cells and synchronizes body functions with periods of light and dark to a circadian ("about a day") rhythm.
   a. The suprachiasmatic nucleus projects to sympathetic neurons, which, in turn, project to the pineal gland.
   b. In the dark, the pineal gland increases its secretory output of melatonin, a sleep-inducing hormone.
   4. The **periventricular nucleus** secretes **growth hormone releasing-hormone** (GHRH), **growth hormone-inhibiting hormone** (GHIH or somatostatin), and **thyroid-releasing hormone** (TRH).
 C. The middle region of the hypothalamus is situated directly above the tuber cinereum and the infundibulum and contains 3 nuclei (Figures 6–2A and B, 6–3).
   1. The arcuate nucleus releases GHRH and prolactin-inhibiting hormone (dopamine). These substances are transported through the hypophyseal portal system into the adenohypophysis.
   2. The **ventromedial nucleus** is a satiety or fullness center.
   3. The **dorsomedial nucleus** is an emotional response center. In experimental animals, stimulation of the dorsomedial nucleus produces **sham rage**, aggressive behavior that lasts only as long as the stimulus.
 D. The **posterior region of the hypothalamic contains the mammillary nuclei in the mammillary bodies** (see Figure 6–2A).
   1. The mammillary nuclei are part of the limbic system.
   2. Other than the infundibulum and neurohypophysis, the mammillary nuclei are the only components of the hypothalamus that are grossly visible on the external surface of the brain.
   a. Most of the input to the mammillary nuclei arises from the subiculum of the hippocampus; these axons reach the mammillary nuclei in the fornix.
   b. The mammillothalamic tract arises from the mammillary nuclei and projects to the anterior nucleus of the thalamus. *Degeneration of mammillary neurons occurs in **Korsakoff's syndrome** and is commonly associated with thiamine deficiency associated with chronic alcoholism. Korsakoff's syndrome is usually preceded by Wernicke's encephalopathy, which includes mental changes, abnormal eye movements, and gait ataxia. In these patients, there are also neuronal changes in the anterior vermis of the cerebellum, in gaze centers*

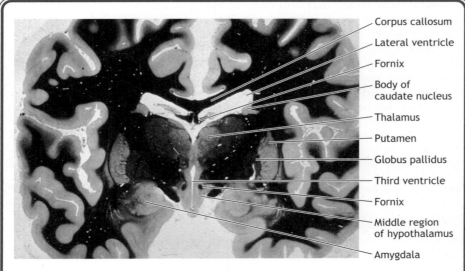

Figure 6–3. Coronal section stained for myelin through the telencephalon and diencephalon.

or ocular neurons in the midbrain and pons, and in the mediodorsal nuclei of the thalamus. Patients with Korsakoff's syndrome have anterograde short-term memory loss, which may be combined with retrograde amnesia and confabulations, in which patients make up stories about events that never occurred.

E. The hypothalamus may also be divided into **anterior and posterior zones**, which control the activity of the parasympathetic and sympathetic divisions of the autonomic nervous system (see Figure 6–2A).

1. The anterior zone mediates parasympathetic effects to conserve energy and restore body resources.

   a. Stimulation of this area promotes a sense of fullness, restfulness, and a reduction in body temperature, which results from sweating and vasodilation.

   b. Parasympathetic effects may include reductions in heart rate, respiration, and blood pressure and an increase in peristalsis.

   c. Sweating, which reduces body temperature, results from central parasympathetic stimulation, which activates eccrine sweat glands innervated by preganglionic and postganglionic cholinergic neurons.

   d. The anterior zone of the hypothalamus is adjacent to the ventral lateral preoptic nucleus, which functions in the initiation of non-REM sleep.

   e. The lateral part of the anterior zone of the hypothalamus contains a feeding center, which promotes feeding when stimulated.

HYPOTHALAMIC LESIONS AND EATING DISORDERS

Lesions of the ventromedial hypothalamus may result in obesity. *Lesions of the lateral hypothalamus* produce severe aphagia.

2. The posterior zone of the hypothalamus mediates sympathetic effects.
   a. Stimulation of this area results in an increase in aggressive behavior and awareness, increased hunger, and an increase in body temperature, which results from shivering and vasoconstriction associated with catabolic activity.
   b. Sympathetic effects may include increases in heart rate, cardiac output, and respiration and decreased activity in gastrointestinal and urogenital viscera.
   c. The posterior zone of the hypothalamus contains histaminergic neurons, which project to cholinergic neurons in the ascending arousal system in the brainstem. These cholinergic neurons function to maintain a state of arousal.

HYPOTHALAMIC LESIONS AND BODY TEMPERATURE

*A patient with a **lesion of the posterior hypothalamus** may have a body temperature that varies with the environmental temperature like a poikilotherm (cold-blooded organism). Conversely, a patient with a **lesion of the anterior hypothalamus** may have hyperthermia.*

3. Descending hypothalamic axons arise from the anterior and posterior zones of the hypothalamus and course through the brainstem and spinal cord to synapse with preganglionic parasympathetic and preganglionic sympathetic neurons (see Figure 6–2A).

F. The **hypothalamus controls endocrine function** by regulating the secretory activity of the pituitary gland; the **pituitary consists of a neurohypophysis and an adenohypophysis**, which have different embryonic origins and different mechanisms of secretion (see Figure 6–2B).

1. The neurohypophysis forms the posterior lobe of the pituitary (see Figure 6–2B).
   a. The neurohypophysis is not a gland; it is a direct outgrowth of the central nervous system (CNS) that contains the axons of large (magnocellular) neurons in the supraoptic and paraventricular nuclei of the hypothalamus.
   b. The axons of the supraoptic and paraventricular nuclei course through the median eminence (part of the tuber cinereum), through the infundibulum in the hypothalamohypophysial tract, and into the posterior lobe of the pituitary.
   c. The posterior lobe consists of axon terminals of supraoptic and paraventricular neurons, which contain vasopressin and oxytocin.
   d. Vasopressin and oxytocin are released into a bed of fenestrated capillaries provided by the inferior hypophyseal artery.
   e. Hypophyseal veins carrying vasopressin and oxytocin drain into the cavernous sinuses, a pair of dural venous sinuses situated adjacent to the pituitary.

2. The adenohypophysis forms the anterior lobe of the pituitary (see Figure 6–2B).
   a. The adenohypophysis is a true endocrine gland that is not derived from the CNS but develops from **Rathke's pouch**, an outgrowth of oral ectoderm.
   b. The secretory activity of the adenohypophysis is regulated by releasing and inhibitory factors produced and secreted by small (parvocellular) neurons in the periventricular and arcuate nuclei.
   c. The releasing and inhibitory factors are transported by axons and released into a primary capillary bed in the median eminence formed from the superior hypophyseal artery.

d. The releasing and inhibitory factors pass from the primary capillary bed through hypophyseal portal veins to reach a secondary capillary bed, from which they are released to influence the secretory activity of the cells in the adenohypophysis.

   3. The adenohypophysis contains a variety of endocrine cells that produce and secrete 3 glycoprotein hormones, 2 mammosomatotropic hormones, and an opiomelanocortin hormone.

    a. The 3 glycoprotein hormones are thyroid-stimulating hormone (TSH), luteinizing hormone (LH), and follicle-stimulating hormone (FSH).

    b. The 2 mammosomatotropic hormones are growth hormone (GH) and prolactin.

    c. The opiomelanocortin hormone is adrenocorticotropic hormone (ACTH).

    d. Hypophyseal veins carrying the secretory products of cells in the adenohypophysis drain into the cavernous sinuses.

IV. The epithalamus is the part of the diencephalon located in the region of the posterior commissure that consists of the pineal gland and the habenular nuclei (see Figure 1–6).

  A. The **pineal gland** is a small, highly vascularized structure.

  B. The pineal gland is situated above the posterior commissure and is attached by a stalk to the roof of the third ventricle.

   1. The pineal gland contains pinealocytes and glial cells but no neurons. Pinealocytes synthesize melatonin, serotonin, and cholecystokinin.

   2. Environmental light regulates the secretory activity of the pineal gland through a retinal-suprachiasmatic-sympathetic-pineal pathway.

   3. Daylight reduces the synthesis and secretion of melatonin by the pineal gland; darkness increases the synthesis and release of melatonin.

   4. Melatonin has an effect on the secretory activity of anterior pituitary hormones, which regulate the activity of the gonads, and has sleep-inducing and hypnotic capabilities.

INSOMNIA AND CALCIFICATION OF THE PINEAL GLAND

The pineal gland is the most common site of calcification in the CNS and may be used as a diagnostic marker in imaging. Pineal calcification may reduce the output of melatonin and be a cause of insomnia.

PINEAL TUMORS AND PARINAUD'S SYNDROME

Pineal tumors *may cause obstruction of cerebrospinal fluid (CSF) flow, increased intracranial pressure, and a noncommunicating hydrocephalus. Compression of the upper midbrain, pretectal area, or posterior commissure by a pineal tumor may cause **Parinaud's syndrome**, in which a patient may have impaired conjugate vertical gaze and pupillary light reflex abnormalities.*

  V. The subthalamic nucleus is functionally associated with the basal ganglia (see Chapter 5).

VI. The limbic system is interconnected with the diencephalon and basal ganglia and functions in homeostasis, olfaction, memory processing, and emotions (mnemonic: HOME).

   A. The **major limbic structures** are the **hippocampus, amygdala, and septal region**, a group of subcortical structures found on the medial aspect of each hemisphere, and the **limbic cortex** in the frontal, parietal, and temporal lobes (Figure 6–4A–C).

   B. The **subcortical limbic structures** project mainly to the mediodorsal and anterior nuclei in the thalamus and to all parts of the hypothalamus.

   C. The **limbic cortex** in the frontal, parietal, and temporal lobes includes the piriform cortex, cingulate gyrus, parahippocampal gyrus, prefrontal and orbitofrontal gyri, and the insula.

VII. The olfactory system is unique among sensory systems in that some of its information can reach the limbic cortex without relaying through a thalamic nucleus and without a significant contralateral projection (see Figure 6–4A).

   A. Central processes of **bipolar olfactory neurons** in the nasal cavity epithelium enter the anterior cranial fossa through the cribriform plate and synapse in the olfactory bulb.

   B. The **olfactory bulb** processes and refines olfactory information.

   C. Axons exit the bulb in the lateral olfactory tract and project directly to several cortical areas and to the amygdala.

   1. Olfactory tract axons project directly to the piriform cortex in the uncus, the entorhinal cortex in the parahippocampal gyrus, and the corticomedial nuclei of the amygdala.

   2. Olfactory tract axons project to the mediodorsal nucleus of the thalamus, which projects to the orbitofrontal cortex.

   3. Olfactory tract axons project to an anterior olfactory nucleus, which then projects to the opposite olfactory bulb by way of the anterior commissure.

OLFACTORY DEFICITS

Olfactory deficits may be incomplete (**hyposmia**), distorted (**dysosmia**), or complete (**anosmia**) and result from transport problems by damage to the primary olfactory neurons or to neurons in the olfactory pathways in the CNS. Head injuries that fracture the cribriform plate may tear the central processes of olfactory nerve fibers as they pass through the plate to terminate in the olfactory bulb or may injure the bulb itself.

VIII. The hippocampus consolidates short-term declarative memories to long-term declarative memories.

   A. Declarative memories are memories of events and facts.

   1. Short-term declarative memories are retained for minutes to hours.

   2. Long-term declarative memories are stored for days to years in adjacent cortical areas.

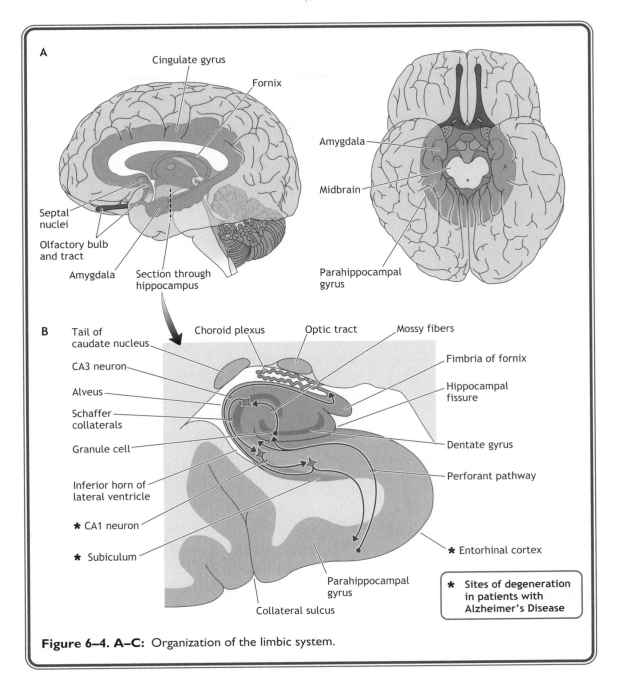

A

Cingulate gyrus

Fornix

Amygdala

Midbrain

Septal nuclei

Olfactory bulb and tract

Amygdala

Section through hippocampus

Parahippocampal gyrus

B

Tail of caudate nucleus

Choroid plexus

Optic tract

Mossy fibers

CA3 neuron

Alveus

Schaffer collaterals

Granule cell

Inferior horn of lateral ventricle

* CA1 neuron

* Subiculum

Fimbria of fornix

Hippocampal fissure

Dentate gyrus

Perforant pathway

* Entorhinal cortex

Parahippocampal gyrus

Collateral sulcus

* **Sites of degeneration in patients with Alzheimer's Disease**

Figure 6–4. A–C: Organization of the limbic system.

B. The **hippocampus** consists of the hippocampus proper, the dentate gyrus, and the subiculum (see Figure 6–4B and C).

        1. Each component of the hippocampus consists of 3 neuron cell layers: a molecular layer, a polymorphic layer, and a middle layer containing pyramidal or granule cells.

        2. The hippocampus proper consists of the cornu ammonis (Ammon's horn), which is subdivided into CA1, CA2, and CA3 regions. These regions contain pyramidal cells; large pyramidal cells are particularly prominent in the middle layer of CA3.

        3. The dentate gyrus contains densely packed granule cells in its middle layer and is interlocked with the hippocampus proper.

        4. The subiculum is a 3-layered cortex in the superior part of the parahippocampal gyrus; the middle layer of the subiculum also contains pyramidal cells.

    C. The **entorhinal cortex**, a 6-layered cortical area in the inferior part of the parahippocampal gyrus, receives most of the input from other cortical areas destined for the hippocampus.

        1. Excitatory (glutamatergic) information flows from the entorhinal cortex through the dentate gyrus, hippocampus proper, and subiculum; information exits the hippocampus mainly by way of axons in the fornix (see Figure 6–4C).

        2. Entorhinal pyramidal cell axons enter the hippocampus in the **perforant pathway** and project to granule cells in the dentate gyrus.

        3. Granule cells in the dentate gyrus project by way of **mossy fibers** to pyramidal cells in the CA2 and CA3 regions of the hippocampus proper.

        4. CA3 axons in the hippocampus proper project to CA1 and project to pyramidal cells in the subiculum by way of **Schaffer collaterals**.

        5. The subiculum projects back to the entorhinal cortex; entorhinal axons also exit the hippocampus in the fornix.

        6. The perforant pathway, mossy fibers, and Schaffer collaterals use glutamate and NMDA (*N*-methyl-D-aspartate) glutamate receptors to facilitate long-term potentiation, plasticity, and learning in the hippocampus.

SEIZURES AND THE HIPPOCAMPUS

The lack of inhibitory neurons in hippocampal structures makes this region more susceptible to seizures than other cortical areas.

ALZHEIMER'S DEMENTIA AND THE LIMBIC SYSTEM

• *The **neural degeneration in patients with Alzheimer's disease** is characterized by an accumulation of neurofibrillary tangles and senile plaques and occurs initially in the entorhinal cortex and then in the subiculum and CA1 region of the hippocampus proper. The basal nuclei of Meynert, limbic nuclei of the thalamus, neurons in association cortical areas, and noradrenergic and serotoninergic neurons in the brainstem are affected later. Primary sensory and motor cortical areas are spared, as are cranial nerve nuclei and the cerebellum.*

• *In dementia of the Alzheimer's type, impaired memory for events in time is often the leading symptom, followed by intellectual decline, language deficits, which may lead to an aphasia, and changes in mood or personality (e.g., depression, paranoia). As the dementia progresses, deficits in orientation, reasoning, judgment, and the ability to abstract, calculate, and use language become evident. Routine activities may be handled reasonably well, but the patient fails to respond as well to new or altered situations. Patients may confabulate and deny or hide their deficits.*

  D. The **Papez circuit of limbic neural connections** most likely participates in memory consolidation (Figure 6–5).

   1. The Papez circuit arbitrarily begins with entorhinal input, which flows through the hippocampus (see prior discussion).
   2. The fornix fibers are axons that leave the subiculum and project mainly to the mammillary nuclei (postcommissural fornix) in the posterior hypothalamus.
   3. Fornix axons in the precommissural fornix project to the septal nuclei and anterior nucleus of the thalamus.
   4. Cholinergic projections from the septal nuclei and basal nucleus of Meynert project back to hippocampal structures and enhance memory consolidation.
   5. The mammillary nuclei project to the anterior nucleus of the thalamus, which projects to the cingulate gyrus.
   6. The cingulate gyrus completes the Papez circuit by projecting back to entorhinal cortex in the parahippocampal gyrus in the cingulum.

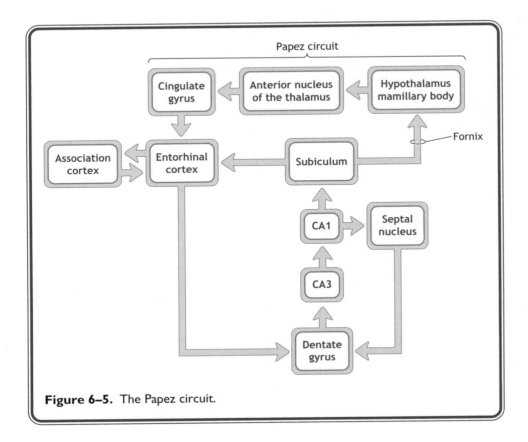

Figure 6–5. The Papez circuit.

IX. The amygdala uses visceral and somatic sensory inputs to coordinate emotional responses to pleasurable, fearful, and visceral stimuli (see Figure 6–4A).

A. The amygdala is situated in the rostral and medial parts of the temporal lobe deep to the uncus.

B. The amygdala receives highly processed visual, auditory, somatosensory, and visceral inputs from wide areas of the cerebral cortex and from the hypothalamus.

C. The amygdala contains basolateral and corticomedial nuclei.

    1. The basolateral nuclei receive mostly inputs from external stimuli by way of the visual, auditory, and somatosensory systems.

        a. The basolateral nuclei contain GABA neurons.

        b. The basolateral nuclei of the amygdala project by way of the ventral amygdalofugal pathway to the mediodorsal nucleus of the thalamus, to the prefrontal and orbitofrontal cortex, and to the hippocampus.

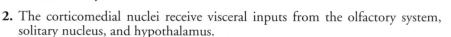

GABA NEURONS

*The **GABA neurons in the basolateral nuclei** are sensitive to GABA agonists, which enhance GABA transmission and reduce anxiety.*

    2. The corticomedial nuclei receive visceral inputs from the olfactory system, solitary nucleus, and hypothalamus.

        a. The corticomedial nuclei project by way of the stria terminalis mainly to the hypothalamus, septal region, and basal nucleus of Meynert.

        b. The corticomedial nuclei function in the evaluation of visceral stimuli associated with reproduction and survival.

        c. Cholinergic neurons in the basal nucleus of Meynert project to wide areas of the cerebral cortex and enhance levels of arousal in response to an emotional stimulus.

KLÜVER-BUCY SYNDROME

Klüver-Bucy syndrome results from bilateral lesions to or ablation of the amygdala and hippocampus. Patients with amygdala lesions in Klüver-Bucy syndrome become placid, exhibiting a marked decrease in aggressive behavior, and show little emotional reaction to external stimuli. These patients may exhibit hypermetamorphosis, in which visual objects are repeatedly approached as though they were completely new. These patients may have increased oral exploratory behavior, in which they put everything in their mouths, eating only appropriate objects, and become hypersexual. If the hippocampus is involved, patients may have anterograde amnesia. Rostral temporal lobe lesions may also affect the lateral part of the visual radiation (Meyer's loop) and result in a contralateral homonymous superior quadrantanopsia (see Chapter 7).

CLINICAL PROBLEMS

1. Your patient has a tumor in the third ventricle, which is compressing the hypothalamus. The patient exhibits signs associated with a decreased output of vasopressin by the posterior pituitary. What structure has the tumor compressed?

   A. Supraoptic nucleus

   B. Suprachiasmatic nucleus

   C. Arcuate nucleus

   D. Medial preoptic nucleus

   E. Ventromedial nucleus

2. Your patient has been diagnosed with an eating disorder. The patient seems to have lost a significant amount of weight in the last year. A lesion in which of the following areas might have caused the patient's problem?

   A. Medial preoptic nucleus

   B. Paraventricular nucleus

   C. Mammillary body

   D. Ventromedial nucleus of hypothalamus

   E. Lateral part of the anterior zone of hypothalamus

3. A magnetic resonance image of a female patient reveals that a calcified neural structure is compressing structures adjacent to the superior colliculus. What complaint might you expect the patient to have?

   A. Always being hot regardless of the environmental temperature

   B. Insomnia

   C. Amnesia

   D. Hearing voices telling wildly untrue stories

   E. Difficulty remembering where she put the car keys each morning

4. On New Year's Eve, an elderly male staggers into the emergency room smelling of alcohol, unshaven, and having no wallet or identification. The patient gave his name but did not remember his address or any other personal information. The patient said he was on a business trip to the city but could not recall where he was staying. When questioned 5 minutes later, he could not remember why he was in the city, claimed he was an off-duty airline pilot, and identified one of the nurses as a flight attendant. A neurological exam revealed a gaze-evoked nystagmus in all directions. His muscle strength was 5/5, but he walked with a broad-based gait. Together, these signs and symptoms suggest that the patient may have:

   A. Alzheimer's disease

   B. Subacute combined degeneration

   C. Huntington's disease

D. Korsakoff's syndrome

E. Neurosyphilis

5. What might be one site of neuronal degeneration in the patient in the previous description?

A. Head of the caudate nucleus

B. Substantia nigra, pars compacta

C. Mammillary bodies

D. Ventrobasal complex

E. Basal nucleus of Meynert

MATCHING PROBLEMS

Questions 6–15: Nuclei match

Use the following diencephalic nuclei to answer the questions below:

Choices (each choice may be used once, more than once, or not at all):

A. Medial dorsal nucleus

B. Ventral anterior nucleus

C. Mammillary nuclei

D. Ventral posterior medial nucleus

E. Suprachiasmatic nucleus

F. Supraoptic nucleus

G. Arcuate nucleus

H. Lateral zone of hypothalamus

I. Ventromedial nucleus of hypothalamus

J. Pulvinar

6. Lesions here result in aphagia

7. Helps regulate circadian rhythms

8. Lesions here result in obesity

9. Relays taste information to cortex

10. Relays somatosensory information from the face to cortex

11. Produces dopamine

12. Sends axons to the posterior pituitary

13. Produces oxytocin

14. Part of the Papez circuit

15. Site of synapse of globus pallidus axons

Questions 16–25: Nuclei match 2

Use the following diencephalic nuclei to answer the questions below:

Choices (each choice may be used once, more than once, or not at all):

    A. Anterior nucleus

    B. Medial dorsal nucleus

    C. Medial geniculate nucleus

    D. Lateral geniculate nucleus

    E. Ventral lateral nucleus

    F. Ventral posterior medial nucleus

    G. Ventral posterior lateral nucleus

    H. Ventral anterior nucleus

16. Projects to Brodmann areas 41 and 42

17. Projects to prefrontal cortex

18. Receives input from the dentate nucleus of the cerebellum

19. Projects to cingulate gyrus

20. Retinal ganglion cells synapse here

21. Spinothalamic axons synapse here

22. Projects to Brodmann area 17

23. Projects to same cortical area as the lateral dorsal nucleus

24. Receives input from both basal ganglia and cerebellum

25. Receives input from the trigeminal nuclei

ANSWERS

1. The answer is A. Vasopressin is produced and secreted by the supraoptic nucleus and the paraventricular nucleus.

2. The answer is E. Lesions of the lateral hypothalamus produce severe aphagia.

3. The answer is B. The patient has a calcified pineal gland, which may reduce the output of melatonin, a sleep-inducing hormone.

4. The answer is D. Korsakoff's syndrome is usually preceded by Wernicke's encephalopathy, which includes mental changes, abnormal eye movements, and gait ataxia. Patients with Korsakoff's syndrome have anterograde short-term memory loss, which may be combined with confabulations, in which they make up stories about events that never occurred.

5. The answer is C. In these patients, there are neuronal changes in the anterior vermis of the cerebellum, in gaze centers or ocular neurons in the midbrain and pons, and in the mammillary bodies and dorsomedial nucleus of the thalamus.

6.	H	**13.**	F	**20.**	D
7.	E	**14.**	C	**21.**	G
8.	I	**15.**	B	**22.**	D
9.	D	**16.**	C	**23.**	A
10.	D	**17.**	B	**24.**	E
11.	G	**18.**	E	**25.**	F
12.	F	**19.**	A		

CHAPTER 7

VISUAL SYSTEM, PUPILLARY REFLEXES, AND CONJUGATE EYE MOVEMENTS

I. **The visual system is the sensory system that is most relied on by humans. An estimated 40% of the neurons in the central nervous system (CNS) contribute to visual function. Many neurological disorders are frequently associated with some type of visual dysfunction.**

   A. In the visual system, the **retina contains the photoreceptors that respond to light**. An initial processing of visual information occurs there.

   B. The **visual system uses 3 neurons,** like other major sensory systems, to **convey information from the photoreceptors to the cerebral cortex**. The axons of the second neuron in the visual pathway (the optic nerve) are partially crossed.

   C. **Other regions of the CNS** that receive direct visual inputs include the **suprachiasmatic nucleus** in the hypothalamus, the **pretectal nuclei** in the midbrain, and the **superior colliculus**.

   1. The suprachiasmatic nucleus regulates circadian rhythms by synchronizing body functions with periods of light and dark.

   2. The pretectal nuclei function in the pupillary light reflex, and the superior colliculus functions in conjugate gaze (see later discussion).

II. **The eye is specialized for the collection and refraction of light (Figure 7–1A).**

   A. Light enters the eye through 2 transparent structures: the **cornea** and the **lens**. These structures **refract light so that it impinges on the retina**.

   1. The cornea is the slightly curved avascular outer coating of the eye and is the fixed principal refractive medium of the eye.

HYPEROPIA, MYOPIA, AND ASTIGMATISM

- *Hyperopia,* or farsightedness, results from a flat cornea that has too little refractive power and focuses an object behind the retina.
- Myopia, or nearsightedness, results from a cornea that is too round, has too much refractive power, and focuses an object in front of the retina.
- *Astigmatism* results from an irregularly shaped cornea that transmits distorted images.

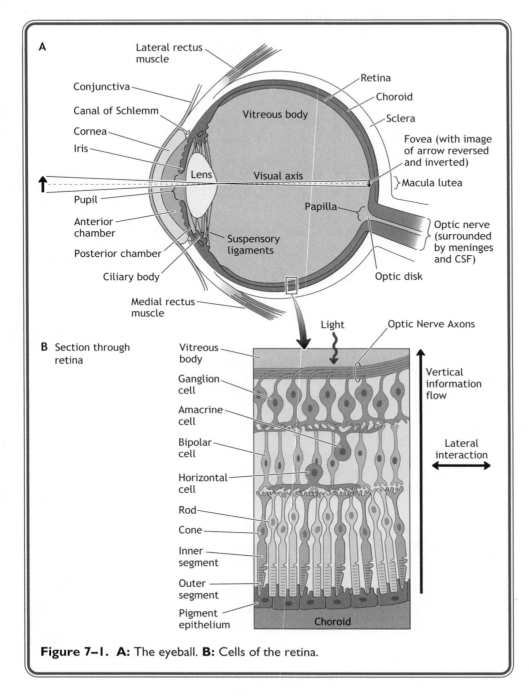

Figure 7–1. A: The eyeball. **B:** Cells of the retina.

2. The **iris** and its circular opening, the **pupil**, are in front of the lens.
   a. The iris contains the 2 smooth muscles, the sphincter and the dilator pupillae, which act to change the diameter of the pupil from 1–8 mm and control the amount of light entering the eye.
   b. The dilator pupillae muscle is innervated by preganglionic sympathetic fibers from the T1 segment of the spinal cord and by postganglionic sympathetic axons from the superior cervical ganglion.
   c. The sphincter pupillae muscle is innervated by preganglionic parasympathetic fibers from the Edinger-Westphal nucleus that exit the midbrain in the oculomotor nerve (CN III) and by postganglionic parasympathetic axons from the ciliary ganglion.

HORNER'S SYNDROME

- *Horner's syndrome* results from a lesion of one of the neurons used in the sympathetic innervation of the face, scalp, and orbit. Patients have miosis, ptosis, and anhidrosis on the side of the lesion.
- Horner's syndrome may be caused by a **lesion of preganglionic sympathetic axons** that exit from the T1 segment of the spinal cord, a **lesion of the cervical sympathetic trunk or superior cervical ganglion**, or a **lesion of descending hypothalamic axons** in the CNS.

MYDRIASIS

- *Mydriasis*, a dilated pupil, may be an initial sign in patients with external compression of the oculomotor nerve (CN III) because the preganglionic parasympathetic fibers course in the peripheral part of the nerve.
- **External compression of the oculomotor nerve** may be caused by **uncal herniation** or a **berry aneurysm**.

3. The anterior and posterior chambers are situated between the cornea and lens and are separated by the iris and the pupil.
   a. The anterior and posterior chambers contain aqueous humor, which is produced by the ciliary body in the posterior chamber.
   b. Aqueous humor flows through the pupil and drains through the trabecular meshwork to the canal of Schlemm, a specialized vein near the junction of the iris and the cornea in the anterior chamber.
   c. The lens forms the posterior boundary of the posterior chamber and is held in place by the fibers of the suspensory ligament.
   d. Aqueous humor in the posterior chamber is in contact with the vitreous body behind the lens through the fibers of the suspensory ligament.

AQUEOUS HUMOR, INTRAOCULAR PRESSURE, AND GLAUCOMA

- *Glaucoma* is caused by an **increase in intraocular pressure** that results initially in compression of the longest axons of the optic nerve that arise from the peripheral part of the retina and a loss of peripheral vision. Untreated glaucoma may result in papilledema and may lead to complete blindness.
- **Open-angle glaucoma** is the most common form of glaucoma and results from an obstruction of the canal of Schlemm. **Closed-angle glaucoma** results from an increase in intraocular pressure secondary to an adhesion of the peripheral part of the iris to the cornea that prevents aqueous humor from reaching the trabecular meshwork.

4. The **ciliary body** contains the ciliary muscle.
   a. The ciliary muscle is a smooth muscle that, when contracted, relaxes the suspensory ligaments of the lens, allowing the lens to round up for near vision.
   b. Contraction of the ciliary muscle is part of the accommodation reflex, or near response, under control of parasympathetic and skeletal motor fibers in the oculomotor (CN III) nerve.

PRESBYOPIA

Presbyopia results when the lens becomes less elastic, reducing the ability to focus on near objects. The lens, like the cornea, can develop opacities known as cataracts. Lens replacements restore visual clarity but not accommodation.

B. The **vitreous body contains vitreous humor** and is between the lens and the retina.
   1. The vitreous humor is a thick, gelatinous matrix that contains phagocytic cells, which remove blood and debris from the vitreal matrix.
   2. "Floaters" are accumulations of debris too large for phagocytic removal from the vitreous matrix.

III. **The neural retina consists of the photoreceptors, 4 types of neurons, including projection neurons (bipolar cells and ganglion cells), and interactive neurons (amacrine cells and horizontal cells), plus glial cells (see Figure 7–1B).**

A. The **photoreceptors** of the retina are the **rods and cones**.
   1. The photoreceptors are situated in the external part of the retina so that light must travel through other layers of the retina to reach them.
   2. The outer segments of rods and cones contain disk membranes in which light is detected and transduced into graded membrane potentials.
      a. The outer segments contain photopigments, which consist of 11-*cis* retinal, a derivative of vitamin A that is bound to an opsin protein; rods contain rhodopsin, and cones contain 3 different kinds of iodopsin.
      b. Absorption of a photon of light activates rhodopsin and 11-*cis* retinal is converted to all-*trans* retinal, which activates an intracellular messenger, **transducin**.
      c. Transducin activates a phosphodiesterase that hydrolyzes cGMP (cyclic guanosine monophosphate) that reduces its concentration and closes cGMP-gated channels, allowing the photoreceptor membrane to hyperpolarize.
      d. Under light conditions, Na^+ permeability is decreased, and the amount of glutamate released by the photoreceptors is decreased proportionally.
      e. The all-*trans* retinal is reduced to all-*trans* retinol, which is transported to the pigment epithelium, where it is converted back to 11-*cis* retinal.
   3. There is only 1 kind of rod.
      a. Rods are specialized for achromatic vision at low light levels.
      b. There are about 90 million rods distributed throughout the retina except in the fovea.

c. Rods have high light sensitivity and are specialized for scotopic (night) vision, but have low spatial resolution.

4. There are about 4.5 million cones in the retina; there are 3 types of cones, each with a different photopigment sensitive to a different wavelength of light.

    a. L cones respond to red wavelengths, M cones to green wavelengths, and S cones to blue wavelengths; color information is obtained by comparing the responses from red, green, and blue cones.

    b. Cones are less sensitive to light than rods, are specialized for photopic (daytime) vision, and are for acuity and color vision.

NIGHT BLINDNESS, COLOR BLINDNESS, AND PATIENTS WHO ARE LEGALLY BLIND

- *A total loss of rods results in **night blindness**, and patients with a total loss of cones are legally blind. A significant decrease in dietary vitamin A is a cause of night blindness.*
- ***Color blindness** is more common in males than females because males lack one of the genes that encode for red or green pigments on their single X chromosome. Blue pigments are less likely to be missing because the blue pigment gene is located on chromosome 7.*

    c. Mesopic vision occurs when both rods and cones are activated (starlight and moonlight conditions).

5. Cones are found throughout the retina, but are concentrated in the **foveola**, a rod-free pit in the center of the fovea. Here, the superficial cells of the retina are displaced so that light directly impinges on the cones without being distorted.

    a. The fovea is located on the temporal side of the optic disk, where optic nerve axons exit the retina.

    b. The fovea is surrounded by the macula luteae, a yellow-pigmented rim.

6. The inner segments of rods and cones contain mitochondria and the nucleus and synapse with bipolar cells and horizontal cells.

B. Rods and cones synapse with two types of bipolar cells that are the first neurons in the 3-neuron visual processing pathway that conveys information from the photoreceptors to the cerebral cortex.

1. Hyperpolarizing bipolar cells (OFF cells) are depolarized in the dark and hyperpolarize in response to light.

2. Depolarizing bipolars (ON cells) are hyperpolarized in the dark, and depolarized to light.

3. Rod bipolars exist only in the ON form and are contacted by many rods; this convergence makes the rod system good detectors of light.

4. In most of the retina, fewer cones converge on either ON or OFF cone bipolars.

5. In the fovea, a single cone drives one ON and one OFF cone bipolar, and each cone bipolar is driven by only a single cone; this one to one relationship maximizes acuity.

C. Ganglion cells are the second neurons in the visual pathway and are found in the ganglion cell layer near the vitreal surface of the retina.

1. In most parts of the retina, both rod and cone signals may influence the same ganglion cells; ganglion cells are the only retinal neurons that generate action potentials.
   a. Midget or P ganglion cells are the most numerous ganglion cells (80%); they receive inputs from a single bipolar cell in the fovea, and a few bipolars outside of the fovea, and are specialized for spatial resolution, and color vision.
   b. Parasol or M ganglion cells (10%) receive inputs from many bipolars and are sensitive to movement rather than to resolution and color.
   c. The remaining 10% of retinal ganglion cells project to the suprachiasmatic nucleus of the hypothalamus or to the pretectal area to contribute to the pupillary light reflex.
2. At the optic disk, axons of ganglion cells converge to form the optic nerve; ganglion cells are the only cells in the retina that generate action potentials.
   a. There are no photoreceptors at the optic disk; this area is the blind spot of the eye.
   b. At the disk, optic nerve axons penetrate the choroid and sclera of the eyeball, and the axons acquire a myelin sheath that is formed entirely by oligodendrocytes.
   c. The axons that form the optic nerve are ensheathed by the 3 layers of meninges that surround the rest of the CNS.
   d. The subarachnoid space containing cerebrospinal fluid (CSF) extends along the optic nerve to the optic disk.
   e. The central artery and vein of the retina pass through the optic disk.

INCREASES IN INTRACRANIAL CSF PRESSURE

Increases in intracranial CSF pressure at the optic disk may **reduce axoplasmic flow** in the optic nerve fibers and cause **papilledema**, a swelling of the nerve at the disk. Papilledema results from elevated CSF pressure that is transferred to CSF along the optic nerve and interferes with retinal venous return. Retinal veins become engorged, the optic disk swells, and its edges become less well defined. Flame-like hemorrhages may appear outside the edges of the disk, and fluid may accumulate at the macula.

   D. The **horizontal and amacrine cells** promote lateral interaction within the retina.
      1. Horizontal cells are inhibitory neurons that receive input from rods and cones and enhance luminance contrast by inhibiting adjacent rods and cones.
      2. Amacrine cells are found in both the outer nuclear layer and the ganglion cell layer and synapse with both bipolar cells and ganglion cells.
      3. Amacrine cells use a variety of neurotransmitters and function in the detection of motion, the speed or direction of a moving stimulus, and changes in the intensity of a light stimulus.
   E. **Müller cells** are the glial cells of the retina and function similar to astrocytes in taking up neurotransmitter and ions from the extracellular space.

IV. The pigment epithelium contains melanin and is situated between the choroid and the photoreceptors of the retina.

   A. The pigment epithelium **transports nutrients** (glucose) from the choroidal blood vessels to the photoreceptors.

B. The pigment epithelium contains melanin, which **protects the photoreceptors** by absorbing excessive or scattered light.

C. The pigment epithelium **helps maintain the photoreceptors** by phagocytosing and rapidly recycling the disks in their outer segments.

D. The pigment epithelium **takes up vitamin A** from the blood and stores it as 11-*cis* retinal until it is used by the photoreceptors.

E. The pigment epithelium **takes up all-*trans* retinol** from the outer segments of rods and cones and converts it back to 11-*cis* retinal.

DETACHED RETINA

*A **detached retina** results when there is a separation of the pigment epithelium from the outer segments of the photoreceptors. A detached retina is commonly caused by trauma and may result in degeneration of photoreceptors in the detached segment of the retina.*

AGE-RELATED MACULAR DEGENERATION AND THE PIGMENT EPITHELIUM

Age-related macular degeneration *(AMD) is the most common cause of loss of vision in the elderly. It results from gradual disappearance of the pigment epithelium and then a loss of photoreceptors in the area of the macula.*

RETINITIS PIGMENTOSA

Retinitis pigmentosa *(RP) is a group of hereditary disorders in which there is a progressive visual loss caused by a degeneration of photoreceptors. Rods in the peripheral part of the retina degenerate initially, resulting in a loss of night vision and a loss of peripheral vision. Melanin migrates into the pigment epithelium adjacent to the affected parts of the retina, producing clumps of pigment near retinal blood vessels.*

V. The main components of the visual pathway—the optic nerves, the optic chiasm, and optic tracts (Figure 7–2A)—are different names for axons of retinal ganglion cells.

 A. Each **optic nerve** leaves the eyeball and courses posteromedially through the optic canal at the posterior aspect of the orbit.

 B. The optic nerves converge in the midline just above the pituitary gland to form the optic chiasm and then diverge again.

 C. At the chiasm, there is a partial crossing or decussation of optic nerve fibers; 60% of the fibers from each nerve cross to the opposite side, whereas the other 40% remain uncrossed.

  1. The optic nerve fibers that cross in the chiasm arise from the nasal half of each retina but represent visual information entering the eye from the temporal half of each visual field.

  2. Uncrossed optic nerve fibers arise from the temporal half of each retina and represent visual information from each nasal hemifield.

 D. The **optic tracts** contain remixed (partially crossed) optic nerve fibers that extend from the chiasm.

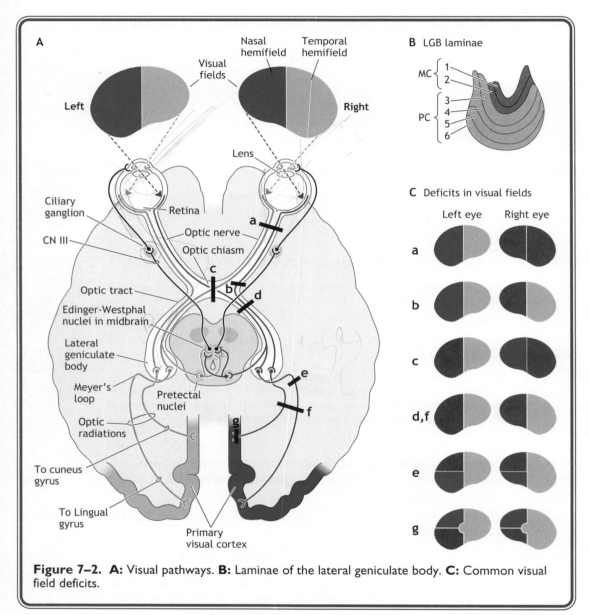

Figure 7–2. A: Visual pathways. **B:** Laminae of the lateral geniculate body. **C:** Common visual field deficits.

1. Most axons of the ganglion cells in the optic tracts project to the lateral geniculate nucleus in the thalamus (see Figure 7–2B).
2. Axons of the ganglion cells of the retina also project to the suprachiasmatic nucleus in the hypothalamus (for circadian rhythms), the pretectal nuclei in the midbrain (for pupillary reflexes), and the superior colliculus (for conjugate gaze).

LESIONS OF AN OPTIC NERVE AND VISUAL FIELD DEFICITS

Lesions of an optic nerve, the optic chiasm, or an optic tract *(see Figure 7–2C) result in deficits that are reversed and inverted in a part of a visual field. A **lesion in part of the visual pathway** originating in the nasal retina will result in deficits in the temporal visual field. A **lesion in the temporal retina or in fibers of the visual pathways** from the temporal retina will result in a deficit in a nasal visual field.*

*Visual field deficits are referred to as **anopsias** or anopias. Losses in half of a field are **hemianopsias**, and losses in a quarter of a visual field are **quadrantanopsias**.*

LESIONS IN FRONT OF THE OPTIC CHIASM

Lesions in front of the optic chiasm *result in visual field deficits that are monocular and ipsilateral to the side of the lesion.*

- *Lesions in the retina produce **scotomas**, small spotlike deficits in the visual field of that eye.*
- *Complete lesions of an optic nerve may produce an **anopsia**, or ipsilateral blindness of that eye, and may be caused by papilledema resulting from increased intracranial CSF pressure or by a blockage of the central artery of the retina. Optic neuritis is common in patients with multiple sclerosis.*
- *An **aneurysm of the internal carotid artery** may compress axons of the optic nerve that arise from the temporal part of the retina. Compression of these axons results in an **ipsilateral nasal hemianopsia** (i.e., compression of temporal retinal axons on the right results in a right nasal hemianopsia).*

LESIONS AT THE OPTIC CHIASM

*A **complete lesion of the fibers that cross at the optic chiasm** results in a **bitemporal heteronymous hemianopsia**, a loss of vision from both the right and left temporal fields. The deficits are heteronymous because the losses are in the temporal parts of the left and right hemifields.*

- *Bitemporal deficits may result from a pituitary adenoma that compresses the chiasm from below. Initially, these patients may have a bitemporal superior quadrantanopsia because axons conveying visual information from superior temporal quadrants cross inferiorly in the chiasm.*
- *Bitemporal deficits may result from a craniopharyngioma (Figure 7–3) or an aneurysm at the junction of an anterior cerebral artery and the anterior communicating artery that compresses the chiasm from above. Initially, these patients may have a bitemporal inferior quadrantanopsia because axons conveying visual information from inferior temporal quadrants of each visual field cross superiorly in the chiasm.*

LESIONS PAST THE CHIASM

- ***Lesions in visual pathways past the chiasm**, including the optic tract and visual radiations, result in visual field deficits that are contralateral, homonymous, and binocular. Homonymous deficits occur in the same (right or left) region of corresponding part of a nasal and temporal visual hemifield. A complete lesion of the left optic tract results in a right (contralateral) homonymous hemianopsia (see Figure 7–2C).*
- *Contralateral homonymous hemianopsias may be caused by an occlusion of the anterior choroidal artery, the main blood supply to the optic tract, or the thalamogeniculate artery, a branch of the posterior cerebral artery that supplies the lateral geniculate nucleus.*

VI. Ocular reflexes include pupillary light reflex and accommodation convergence reflex.

    A. The **pupillary light reflex** results in a direct, consensual pupillary constriction in response to light (see Figure 7–2A).

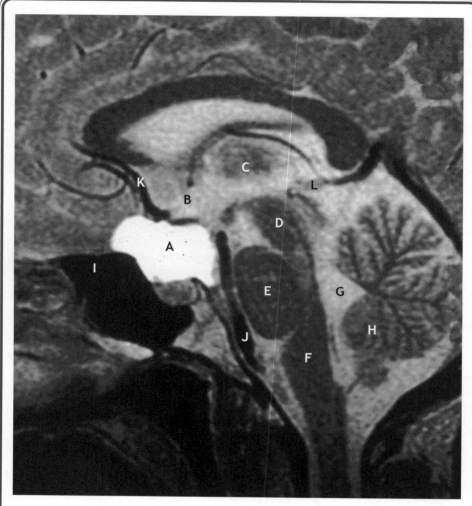

Figure 7–3. T-2 weighted MRI of medial view of the CNS showing a calcified craniopharyngioma at **(A)**. **B:** Hypothalamus. **C:** Thalamus. **D:** Midbrain. **E:** Pons. **F:** Medulla. **G:** Fourth ventricle. **H:** Vermis of cerebellum. **I:** Sphenoid sinus. **J:** Basilar artery. **K:** Anterior cerebral artery. **L:** Pineal gland.

1. The pupillary light reflexes use 4 neurons: the ganglion cell axons of the optic nerve, chiasm, and tract; neurons in the pretectal area; preganglionic parasympathetic fibers of the oculomotor nerve; and postganglionic parasympathetic axons of the ciliary ganglion.
2. In the sensory limb of the reflex, visual information travels from the retina through the optic nerve, chiasm, and tract to synapse in the pretectal nuclei, which are just rostral to the superior colliculi in the midbrain.

3. Axons from the pretectal nuclei cross in the posterior commissure and project bilaterally to preganglionic parasympathetic neurons in Edinger-Westphal nuclei.

4. Each Edinger-Westphal nucleus sends preganglionic parasympathetic axons out of the brainstem in the oculomotor nerve to synapse in the ciliary ganglion in the orbit. Postganglionic axons leave the ganglion to enter the eyeball and innervate the pupillary constrictor muscle in each eye.

5. Shining a light into an eye causes the pupil of that eye to constrict (the direct light reflex) and also causes constriction of the pupil in the other eye, which has not been directly stimulated by light (the consensual light reflex).

B. The **accommodation convergence reflex**, or near response, uses visual input, skeletal motor, and parasympathetic neurons of the oculomotor nerve and ciliary ganglion.

1. When the eyes shift from a far object to a near object, Edinger-Westphal neurons and the ciliary ganglion cause the ciliary muscle to contract, which relaxes the suspensory ligament and increases the curvature of the lens to bring the near object into focus.

2. Both medial rectus muscles contract, causing both eyes to converge on the near stimulus.

3. The pupil constricts, thereby increasing the depth of field of a near object.

ARGYLL ROBERTSON PUPILS

Argyll Robertson pupils are pupils that are **unreactive to light but constrict in the near response**. Argyll Robertson pupils may be seen in patients with tabes dorsalis (caused by neurosyphilis) or diabetes mellitus.

A MARCUS GUNN PUPIL

- *A **Marcus Gunn pupil**, or relative afferent pupillary defect, results from a lesion of an optic nerve and is seen in patients with optic neuritis, which commonly results from multiple sclerosis. A relative afferent pupillary defect may be confirmed by the swinging flashlight test. When light is presented to the normal eye both pupils constrict, but when the flashlight is swung to the affected eye, both pupils paradoxically dilate.*
- *A lesion of an optic tract may result in a slight suppression of the pupillary light reflex; lesions at the level of the lateral geniculate nucleus or in the visual radiations result in a contralateral homonymous hemianopsia with no change in the pupillary light reflex.*

AN EFFERENT PUPILLARY DEFECT

*An **efferent pupillary defect** results from a lesion to the oculomotor nerve. The pupil is dilated on the affected side and will not constrict in response to light presented to either eye or in the near response.*

VII. The lateral geniculate nucleus contains 6 neural layers from which visual radiations arise.

A. The 6 neural layers are numbered 1–6 from ventral to dorsal (see Figure 7–2B).

B. Layers 1 and 2 are magnocellular layers that synapse with optic tract axons from M ganglion cells.

C. Layers 3 through 6 are parvocellular layers that synapse with axons from P ganglion cells.

D. Each geniculate layer also receives nonoverlapping input from either the ipsilateral or contralateral retina.

   1. Ganglion cell axons from the ipsilateral retina that do not cross in the chiasm terminate in layers 2, 3, and 5.

   2. Ganglion cell axons that arise from the contralateral retina cross in the optic chiasm and terminate in layers 1, 4, and 6.

E. The **visual radiations project to the primary visual cortex** (area 17 in the occipital lobe) (see Figure 7–2A).

   1. The medial and superior parts of the visual radiations convey information from the lower quadrants of the visual hemifields and project to the cuneus, the primary visual cortex superior to the calcarine sulcus.

   2. The lateral and inferior parts of the visual radiations convey information from the upper quadrants of the visual hemifields and project to the lingual gyrus, the primary visual cortex inferior to the calcarine sulcus.

      a. The inferior component of the visual radiations forms Meyer's loop, which courses ventrally and rostrally into the temporal lobe.

      b. The fibers of Meyer's loop make a hairpin turn in the temporal lobe before coursing posteriorly through the parietal lobe to reach the lingual gyrus.

LESIONS OF MEYER'S LOOP VERSUS LESIONS OF NON-MEYER'S LOOP COMPONENTS OF THE VISUAL RADIATIONS

CLINICAL CORRELATION

- In the temporal lobe, fibers that form Meyer's loop are supplied by the inferior division of the middle cerebral artery (MCA). (**Meyer's loop** is formed by **lateral** fibers from the **lower** retinal quadrants that project to the **lingual** gyrus.) A stroke involving the inferior division of the MCA that affects Meyer's loop results in a contralateral homonymous superior quadrantanopsia (see Figure 7–2C).
- The non-Meyer's loop fibers are supplied by the posterior cerebral artery. A lesion of this component of the visual radiations results in a contralateral homonymous inferior quadrantanopsia.

VIII. The primary visual or calcarine cortex (V1 or Brodmann area 17), like most other primary sensory cortical areas, is a 6-layered granular cortex.

A. Layer IV is particularly prominent in visual cortex because it is the site of termination of the visual radiation fibers from the lateral geniculate nucleus.

B. Axons of the visual radiations that synapse in layer IV form a distinct band known as the **line of Gennari**.

C. In the primary visual cortex, the peripheral parts of the retina are represented rostrally, and the central part of the retina, including the macula and fovea, is represented close to the occipital pole (Figure 7–4A and B).

D. In the visual cortex, the center surround feature of visual receptive fields are maintained in vertically oriented cortical columns.

   1. Neurons found in all parts of a particular cortical column respond to light stimuli from the same point in a visual field so that the retinotopic order seen at all levels of the visual system is preserved.

   2. Layer IV contains simple cells with small receptive fields that respond best to stationary bars of light with a specific orientation instead of rings or spots.

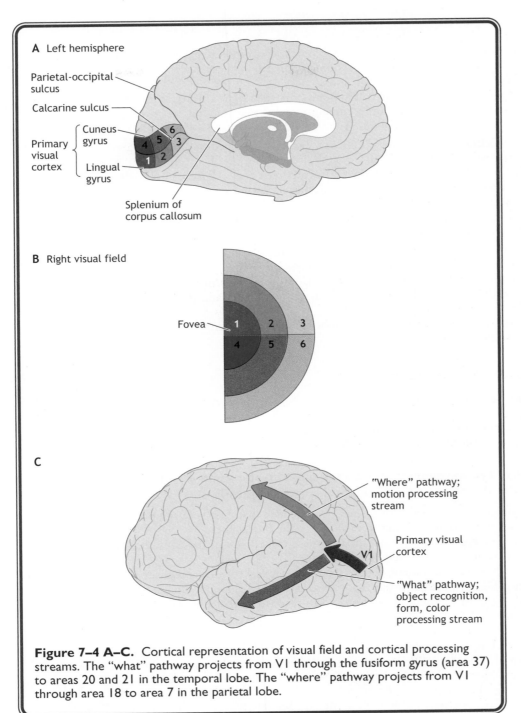

A Left hemisphere

Parietal-occipital sulcus

Calcarine sulcus

Cuneus gyrus

Primary visual cortex

Lingual gyrus

Splenium of corpus callosum

B Right visual field

Fovea

C

"Where" pathway; motion processing stream

Primary visual cortex

V1

"What" pathway; object recognition, form, color processing stream

Figure 7–4 A–C. Cortical representation of visual field and cortical processing streams. The "what" pathway projects from V1 through the fusiform gyrus (area 37) to areas 20 and 21 in the temporal lobe. The "where" pathway projects from V1 through area 18 to area 7 in the parietal lobe.

3. The simple cells in layer IV of each cortical column respond best to input from the right eye or the left eye.

4. Neurons in layers above and below layer IV are complex cells with large receptive fields that respond to bars of light that have a particular orientation and are moving in a specific direction.

5. Stripes forming orientation columns are situated at right angles to the cortical columns. Orientation columns contain neurons that respond best to bars of light with a particular orientation.

E. **A parvocellular stream** contributes to a "what" visual pathway that processes form and color (see Figure 7–4C).

1. The parvocellular stream originates mainly from cones in the central part of the retina, relays through parvocellular layers of the lateral geniculate, and projects to blob/interblob zones of primary visual cortex.

2. These areas project to the inferior part of the temporal lobe to Brodmann areas 37, 20, and 21(see Figure 7–4C).

F. **A magnocellular stream** contributes to a "where" visual pathway that processes motion, depth, and spatial information (see Figure 7–4C).

1. The magnocellular stream originates mainly from rods in the retina, relays through magnocellular layers of the lateral geniculate, and projects to non-blob zones of primary visual cortex.

2. These areas project to the parietal lobe through Brodmann area 18 to area 7 (see Figure 7–4C).

LESIONS IN PRIMARY VISUAL CORTEX

Lesions in primary visual cortex caused by a stroke involving the posterior cerebral artery result in contralateral homonymous hemianopsias but with "macular sparing" in the central part of the visual field. The cones-only area of the fovea in the macula is spared in cortex lesions because the macular cortex has a dual blood supply from both the posterior and middle cerebral arteries.

IX. **Conjugate eye movements (both eyes move in the same direction) use neurons in the brainstem and cerebral cortex (Figure 7–4).**

A. Horizontal conjugate eye movements are generated by neurons in the **paramedian pontine reticular formation** (PPRF) in the caudal pons.

1. The PPRF gives rise to axons that project to neurons in the abducens nucleus, which is contained within the PPRF.

2. Abducens neurons innervate the lateral rectus muscle, which abducts the eye.

3. The PPRF gives rise to axons that course in the medial longitudinal fasciculus (MLF), cross the midline, and project to neurons in the contralateral oculomotor nucleus (see Figure 7–5).

4. Oculomotor neurons innervate muscles (e.g., medial rectus) that adduct the eye during horizontal gaze.

5. Stimulation of the PPRF results in ipsilateral horizontal conjugate gaze (i.e., stimulation of the PPRF on the right results in abduction of the right eye, adduction of the left eye, and horizontal gaze to the right).

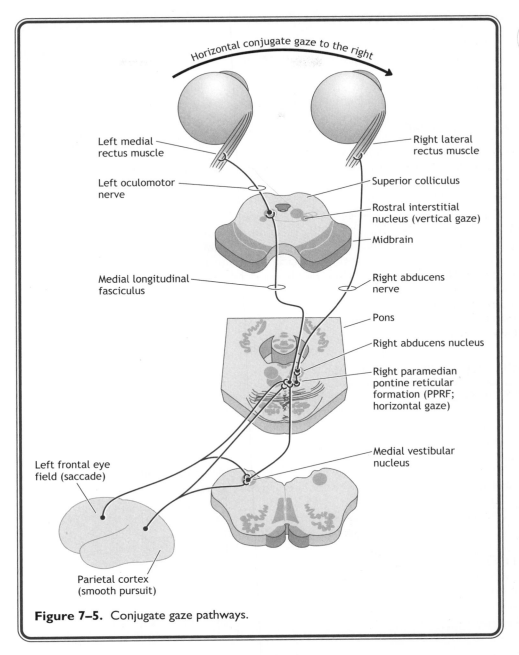

Figure 7–5. Conjugate gaze pathways.

MLF LESIONS AND INTERNUCLEAR OPHTHALMOPLEGIA

- *A **bilateral lesion of the MLF** (commonly seen in patients with multiple sclerosis) results in internuclear ophthalmoplegia in which there is an inability to adduct either eye on attempted conjugate horizontal gaze. Convergence is intact.*
- *In these patients, both eyes abduct normally but may exhibit a monocular nystagmus of the abducting eye. A unilateral lesion of the MLF usually results from an infarction.*

ONE-AND-A-HALF SYNDROME

One-and-a-half syndrome *results from a unilateral lesion of the PPRF, the abducens nucleus, and the axons of the MLF emerging from the PPRF. These patients have an inability to look horizontally toward the side of the lesion with either eye. Horizontal gaze in these patients is limited to an ability to abduct the eye on the side opposite the lesion.*

    B. Vertical conjugate eye movements are generated by neurons in the rostral interstitial nucleus in the midbrain just rostral to the superior colliculus (see Figure 7–5).

        1. The rostral interstitial nucleus gives rise to axons that mainly project a short distance to neurons in the oculomotor nuclei, which innervate muscles that act to elevate (superior rectus and inferior oblique) or depress (inferior rectus) the eye.

        2. The rostral interstitial nucleus also gives rise to axons that course in the MLF to the trochlear nuclei, which innervate the superior oblique muscles that act to depress the eye.

    C. **Saccades** are rapid ballistic voluntary conjugate eye movements (see Figure 7–5) that voluntarily or reflexively activate the gaze centers.

        1. Voluntary saccades do not require a target to be generated, and they function in the dark.

        2. The frontal eye field generates contralateral **horizontal saccades** by activating neurons in the contralateral PPRF or in the deep layers of the superior colliculus, which then project to the contralateral PPRF.

            a. In a horizontal saccade to the right, for example, the left frontal eye field activates the right PPRF, and the right PPRF activates the right abducens nucleus and activates the left oculomotor nucleus through axons of the MLF.

            b. The frontal eye field facilitates the generation of saccades by activating an oculomotor direct basal ganglia pathway (see Chapter 5).

            c. Neurons in the medial vestibular nucleus function to maintain the position of the eyes after a saccade; without these neurons, the eyes would drift after a saccade locates a target.

        3. The frontal eye field generates vertical saccades by activating neurons in the rostral interstitial nucleus.

        4. Reflex saccades are generated by neurons in the deep layers of the superior colliculus, which project to the rostral interstitial nucleus and to the PPRF.

            a. Reflex saccades occur in response to visual, auditory, or somatic stimuli that use these sensory systems to activate the superficial layers of the superior colliculus.

            b. Neurons in the superficial layer activate neurons in the deep layers of the superior colliculus that project to the PPRF and the rostral interstitial nucleus to generate reflex vertical or horizontal saccades.

LESIONS IN THE FRONTAL EYE FIELD OR SUPERIOR COLLICULUS AND THE GENERATION OF SACCADES

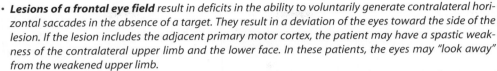

- *Lesions of a frontal eye field* result in deficits in the ability to voluntarily generate contralateral horizontal saccades in the absence of a target. They result in a deviation of the eyes toward the side of the lesion. If the lesion includes the adjacent primary motor cortex, the patient may have a spastic weakness of the contralateral upper limb and the lower face. In these patients, the eyes may "look away" from the weakened upper limb.
- Lesions of the superior colliculus result in transient changes in the accuracy, velocity, and frequency of saccades. The frontal eye field is still able to generate saccades, and the deficits usually improve with time.

    D. Smooth pursuit conjugate eye movements are slow-tracking movements that keep the eyes focused on an object of interest when the object moves across the visual field.

      1. Smooth pursuit movements require a target and do not function in the dark.

      2. Smooth pursuit movements are initiated by neurons in the rod stream in the parietal lobe, which respond to stimuli moving across the visual field.

        a. Neurons in the parietal lobe project to brainstem gaze centers and to the cerebellum.

        b. Neurons in the vermis and flocculus of the cerebellum contribute to the execution of smooth pursuit movements.

      3. The smooth pursuit system may be tested by generating an **optokinetic nystagmus**.

        a. The patient is seated in front of a screen and asked to look at a series of vertical stripes or bars moving across the screen.

        b. The patient's eyes will use smooth pursuit to reflexively follow vertical stripes across the screen. A saccade will then move the eyes in the opposite direction to the other side of the screen.

        c. The alternating smooth pursuit movements in one direction and the saccade generated in the opposite direction in response to a moving stimulus is an optokinetic nystagmus.

BILATERAL LESIONS OF THE MEDIAL PONS AND THE "LOCKED-IN" STATE

The **locked-in state** results from an occlusion of the basilar artery between the branch points of the anterior inferior cerebellar and the superior cerebellar arteries.

- These patients have bilateral lesions of the corticospinal tracts, resulting in a spastic quadriplegia.
- These patients may also have bilateral corticobulbar lesions, resulting in weakness of all cranial-nerve innervated muscles except those served by the oculomotor nerves.
- Body and face sensations are intact, hearing is preserved, and the ascending arousal system is unaffected.
- These patients are fully awake and aware but can communicate only by attempting to blink by inhibiting the levator palpebrae muscles, and moving their eyes vertically.
- Cold-water caloric testing causes a horizontal deviation of the eyes toward the stimulus, but there is no fast or corrective phase of a nystagmus because the frontal eye fields that generate the fast phase are disconnected from the PPRF.

CLINICAL PROBLEMS

1. Your patient has an inability to voluntarily look to the right using horizontal conjugate gaze. What region of the CNS might be a site of the lesion?

   A. Brodmann area 17

   B. Rostral interstitial nucleus on the right

   C. Right frontal eye field

   D. Medial longitudinal fasciculus

   E. Right PPRF

2. Your patient develops an inability to adduct either eye during horizontal gaze. Convergence is intact, and the pupillary light reflex is normal bilaterally. Where is the lesion?

   A. Frontal eye field

   B. Oculomotor nerve

   C. Abducens nerve

   D. Pretectal nuclei

   E. Medial longitudinal fasciculus

3. During an ophthalmic examination, your patient is determined to have elevated intraocular pressure, which might lead to open-angle glaucoma if not treated. What might be the cause?

   A. Lack of transport of fluid from the retina by the pigment epithelium

   B. An excess of vitreous humor

   C. Tumor in the epidural space

   D. Excessive tension on the suspensory ligament

   E. Lack of drainage of aqueous humor into the canal of Schlemm

4. Your patient has a visual field deficit caused by a complete lesion of an optic tract. Which of the following terms may be used to characterize the deficit?

   A. Homonymous

   B. Bitemporal

   C. Ipsilateral

   D. Monocular

   E. Quadrantanopsia

5. Your patient has a complete lesion of the right optic nerve. Where might there be anterograde degeneration of the affected axon terminals?

   A. In laminae 2, 3, and 5 of the left lateral geniculate nucleus

   B. In the ganglion cell layer of the retina

   C. In the supraoptic nucleus of the hypothalamus

D. In laminae 1, 4, and 6 of the left lateral geniculate nucleus

    E. In the Edinger-Westphal nuclei

6. A 41-year-old man had a violent headache and was found unresponsive. He had been known to be hypertensive for the preceding 10 years. He was admitted to a hospital for an evaluation. Both lower limbs were weak and spastic. The upper limbs were also weak, but the patient could move them slightly. Verbal communication was impossible. Horizontal gaze in either direction was abolished, but the patient could look up and down with both eyes, and could partially close both eyes. What might the patient have?

    A. One-and-a-half syndrome

    B. Internuclear ophthalmoplegia

    C. Locked-in state

    D. Medial midbrain syndrome

    E. Parinaud's syndrome

7. A 59-year-old retired autoworker complains of walking into things on his right. He trips over chair legs with his right foot and has trouble driving because he has difficulty seeing cars entering an intersection from the right. A neurological exam reveals a corrected vision of 20/20, normal ocular mobility, and no motor sensory or cranial nerve deficits. Both pupils react briskly to light, and the near response is normal. Visual field testing reveals a right homonymous hemianopsia. Where might the site of a lesion be located?

    A. Optic tract on the left

    B. Meyer's loop on the left

    C. Primary visual cortex in the left occipital lobe

    D. Visual radiations on the left

    E. Frontal eye field on the left

8. A 36-year-old woman who is prematurely amenorrheic is referred to an ophthalmologist because she is having headaches and trouble with her eyes. She cannot see objects off to the side but can see them when they were directly in front of her. A magnetic resonance image reveals that a calcified craniopharyngioma is apparently the cause of the patient's visual problems and amenorrhea. How might you classify her visual field deficit?

    A. Binasal hemianopsia

    B. Contralateral superior quadrantanopsia

    C. Bitemporal inferior quadrantanopsia

    D. Optic scotoma

    E. Bitemporal heteronymous hemianopsia

9. The results of pupillary reflex testing in your patient are shown in Figure 7–6. What might be the cause of their reflex anomaly?

    A. A Marcus Gunn pupil on the right

    B. A berry aneurysm at the junction of the posterior communicating artery and internal carotid artery on the right

    C. Optic neuritis of the right optic nerve

    D. Internuclear ophthalmoplegia

    E. One-and-a-half syndrome

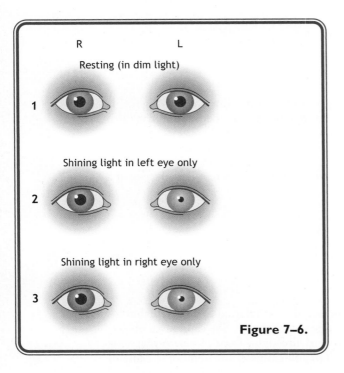

R   L

Resting (in dim light)

1

Shining light in left eye only

2

Shining light in right eye only

3

Figure 7–6.

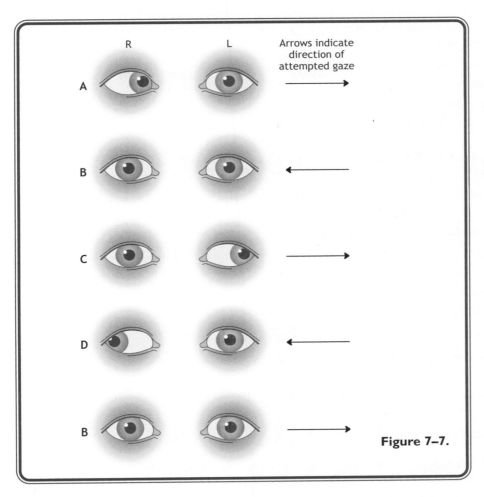

Figure 7–7.

10. Figure 7–7 reveals attempts by a patient to gaze in the direction of the arrow. Of the choices A–E, which one illustrates a patient who has a lesion of the PPRF on the right?

11. Of the choices A–E in Figure 7–7, which one illustrates a patient who has internuclear ophthalmoplegia involving the MLF on the left?

MATCHING PROBLEMS

Questions 12–21: Ocular structures match. Match the choice that is a probable lesion site in each of the patient scenarios that follow.

Choices (each choice may be used once, more than once, or not at all):

A. Left medial longitudinal fasciculus

B. Left oculomotor nucleus

C. Left Edinger-Westphal nucleus

   D. Left PPRF

   E. Left frontal eye field

   F. Right abducens nucleus

   G. Left optic nerve

   H. Left trochlear nerve

   I. Descending hypothalamic axons

12. Patient cannot adduct the left eye during attempted gaze to the right and cannot adduct the left eye in the near response.

13. Patient cannot look horizontally to the right with either eye, and has right lower face weakness.

14. Patient cannot look down with an adducted left eye.

15. When light is presented to the right eye, only the right pupil constricts, when light is presented to the left eye, only the right pupil constricts.

16. Patient has a medially deviated right eye at rest.

17. Patient cannot look horizontally to the left with either eye.

18. Patient cannot look to the right with the left eye; convergence is intact.

19. When light is presented to the right eye, both pupils constrict; when light is presented to the left eye, neither pupil constricts.

20. Patient has a droopy left eyelid, and the diameter of the left pupil is smaller than the right pupil.

21. Patient has an internal strabismus of the right eye.

Questions 22–31: Match each of the following visual field deficits in Figure 7–8 with the appropriate statement.

22. Caused by an aneurysm of the right internal carotid artery

23. Caused by a right temporal lobe tumor

24. Caused by complete compression of the optic chiasm

25. Caused by a lesion of the right optic tract

26. Caused by occlusion of the branches of the posterior cerebral artery to the right cuneus gyrus

27. Caused by lesion to left lingual gyrus

28. Caused by a lesion in the posterior limb of the internal capsule on the right

29. Early manifestation of a pituitary adenoma

30. Caused by a berry aneurysm at the juncture between the anterior cerebral and anterior communicating arteries

31. Early manifestation of a craniopharyngioma

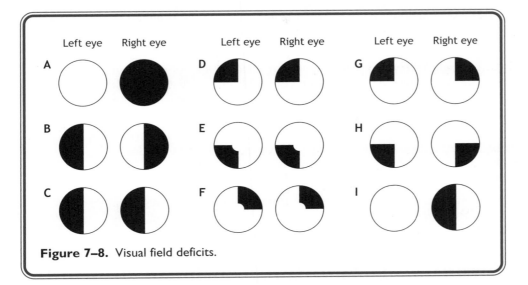

Figure 7–8. Visual field deficits.

ANSWERS

1. The answer is E. The right PPRF is a center for ipsilateral horizontal conjugate gaze. None of the other structures would generate conjugate gaze to the right.

2. The answer is E. The patient may have internuclear ophthalmoplegia, which disrupts the fibers of the MLF between the PPRF and the oculomotor nuclei, which adduct an eye during gaze. Convergence is intact because the near response does not use the MLF.

3. The answer is E. Glaucoma is caused by an increase in intraocular pressure, which results initially in compression of the longest axons of the optic nerve that arise from the perimeter of the retina and a loss of peripheral vision. Open-angle glaucoma is the most common form of glaucoma and results from an obstruction of the canal of Schlemm.

4. The answer is A. Any lesion past the chiasm, including a complete lesion of the optic tract, gives rise to visual field deficits that are homonymous and are in the same part of a temporal and a nasal hemifield.

5. The answer is D. There would be degeneration in laminae 2, 3, and 5 of the right lateral geniculate nucleus. Retrograde changes would be evident in the ganglion cell layer of the retina, and ganglion cells do not project to the supraoptic nucleus of the hypothalamus or directly to the Edinger-Westphal nuclei.

6. The answer is C. This patient has a bilateral lesion of the corticospinal tracts, resulting in a spastic quadriplegia, and a bilateral corticobulbar lesion, resulting in weakness of all cranial nerve innervated muscles except those served by the oculomotor nerves. This patient can communicate only by raising his eyelids and moving his eyes vertically.

7. The answer is D. The light reflex and eye movements are intact so the lesion is not in the optic tract on the left (choice A) or the frontal eye field on the left (choice E). The macula is not spared (choice C), and a lesion at Meyer's loop on the left (choice B) would result in a quadrantanopsia.

8. The answer is C. A craniopharyngioma may compress the superior aspect of the chiasm from above. Initially, these patients may have a bitemporal inferior quadrantanopsia because axons conveying visual information from inferior temporal quadrants of each visual field cross superiorly in the chiasm.

9. The answer is B. The aneurysm has compressed the right oculomotor nerve, resulting in an efferent pupillary defect in which the right pupil does not constrict to light presented to either eye.

10. The answer is B. The patient cannot look to the right with either eye.

11. The answer is D. The patient cannot adduct the left eye during attempted gaze to the right.

12. B	19. G	26. E
13. E	20. I	27. F
14. H	21. F	28. C
15. C	22. I	29. G
16. F	23. D	30. H
17. D	24. B	31. H
18. A	25. C	

CHAPTER 8
CEREBRAL CORTEX AND HIGHER FUNCTIONS

I. The cerebral cortex

A. The **surface of the cerebral cortex** is highly convoluted by gyri that are separated by sulci.

B. On the lateral surface of each hemisphere, the lateral (or Sylvian) fissure and the central sulcus partially subdivide the cortex into **frontal, parietal, temporal, and occipital lobes** (Figure 8–1A).
   1. The **lateral fissure** is a deep sulcus that separates the frontal and temporal lobes rostrally; posteriorly, it partially separates the parietal and temporal lobes.
   2. The **central sulcus** separates the frontal and the parietal lobes, is roughly perpendicular to the lateral fissure, and extends from the superior surface of the hemisphere almost to the lateral fissure.
   3. The boundaries between the parietal, temporal, and occipital lobes on the lateral aspect of the hemisphere are indistinct.

C. On the medial aspect of the hemisphere, the frontal and parietal lobes are separated by a cingulate sulcus from the cingulate gyrus (see Figure 8–1B).
   1. The cingulate gyrus is part of a limbic lobe that includes cortex in the medial parts of the frontal, parietal, and temporal lobes.
   2. Posteriorly, the **parieto-occipital sulcus** separates the parietal lobe from the occipital lobe.
   3. The **calcarine sulcus** divides the occipital lobe into a superior cuneus gyrus and an inferior lingual gyrus. These gyri form the primary visual cortex.

II. Layers of the cerebral cortex

A. There are **6 neuronal cell layers** in 90% of the cerebral cortex; only the piriform (olfactory) cortex, cingulate cortex, and the hippocampal structures consist of fewer than 6 layers (see Figure 8–1C).

B. On the basis of variations in the histology of layers I-VI, Brodmann divided the cortex into 47 areas; many Brodmann numbered areas are commonly used synonymously with functionally specific regions of cortex.
   1. **Layer I, the molecular (plexiform) layer**, is the outermost layer of the cortex and consists primarily of horizontally (tangentially) running nerve fibers.
   2. **Layer II, the external granular layer**, is made up of densely packed granule cells with dendrites extending into the molecular layer and axons passing to deeper cortical layers. Axons of these neurons form association projections that interconnect different parts of a hemisphere.

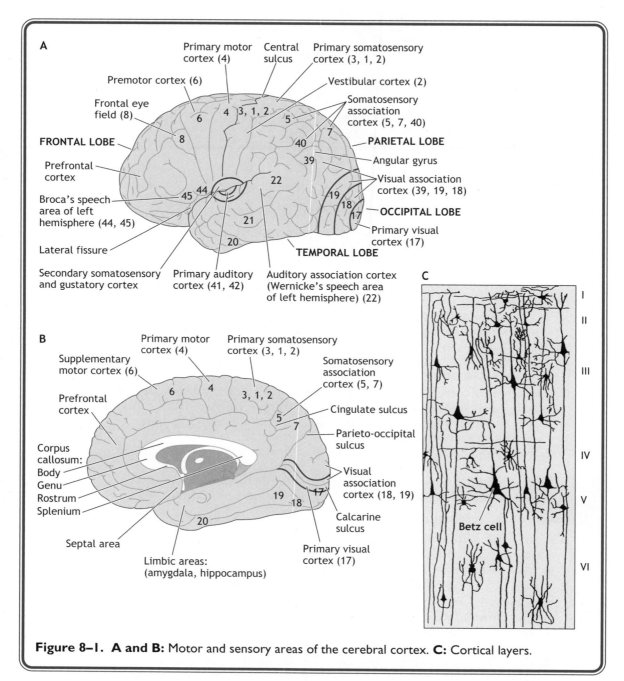

Figure 8–1. A and B: Motor and sensory areas of the cerebral cortex. **C:** Cortical layers.

3. **Layer III, the external pyramidal layer**, consists of moderate-sized pyramidal cells. Axons of these neurons form commissural projections that interconnect the 2 hemispheres via the corpus callosum.

4. **Layer IV, the internal granular layer**, consists mostly of small stellate cells. Layer IV represents the major site of termination of ascending cortical inputs from the thalamus.

5. **Layer V, the internal pyramidal layer**, contains medium to large pyramidal cells.

    a. **Betz cells**, the largest neurons in the CNS, are found in layer V of the primary motor cortex, which is located on the precentral gyrus.

    b. Axons of neurons in layer V form subcortical projections to the basal ganglia and to the cerebellum via the pons.

    c. These axons also form upper motor neurons that project to the brainstem and spinal cord.

6. **Layer VI, the multiform layer**, consists of pyramidal and fusiform cells that vary in size. Axons of these neurons project to the thalamus.

C. In **motor cortical areas**, the pyramidal cell layers III and V are prominent.

D. In **sensory cortical areas**, the granular cell layers II and IV are prominent.

E. In **association cortical areas**, there are 6 layers, but neither the pyramidal nor the granule cell layers are prominent; each functional lobe has a large area of association cortex.

1. Each of the 4 lobes contains a primary cortical area for perception of a specific sensation or a primary motor cortical area involved in the control of skeletal muscles.

2. Each lobe contains an area of association, or "higher function," cortex that is a site of integration of sensory, motor, and behavioral information.

III. Sensory areas of cerebral cortex

A. Sensory areas of cerebral cortex are organized into functional units called **cortical columns;** auditory, visual, and somatosensory cortices contain cortical columns.

B. Each column consists of a **vertical interconnected array of 300–600 neurons**, including cells from layers I–VI.

C. In a sensory cortical column, all neurons may respond to the same modality stimulus from the same location; neurons in an adjacent column may respond to a different modality stimulus but from the same location.

D. In a sensory column, individual axons from a specific thalamic nucleus project to an individual column.

E. Each sensory column acts as a cortical processing unit that is interconnected with adjacent columns.

IV. Hemispheric dominance, handedness, and language

A. Ninety-five percent of the human population is right-handed and has speech and language functions localized in the left or "dominant" hemisphere; the 5% who are left-handed may have speech and language functions localized in the left or right hemisphere.

B. The 2 main language centers are found on the lateral aspect of the left hemisphere in the frontal lobe and in the temporal lobe (see Figure 8–1A).

1. Broca's area is in the inferior part of the frontal lobe anterior and adjacent to the primary motor cortex.

2. Wernicke's area is mainly in the temporal lobe posterior and adjacent to the primary auditory cortex and extends into the parietal lobe.

LESIONS IN CORTICAL AREAS

*Lesions in cortical areas generally result in deficits that begin with the letter A, including **apraxia, agnosia, aphasia, amnesia, alexia, acalculia, abulia, and anopsia**.*

V. Functional features of the frontal, parietal, temporal, and occipital lobes

A. The **frontal lobe** functions in the control of skeletal muscles on the opposite side of the body, contains centers for conjugate eye movements and motor speech, and controls behaviors related to personality and character (Figure 8–2A and B).

1. The precentral gyrus (primary motor cortex) is immediately anterior to the central sulcus and contains upper motor neurons that control skeletal muscles on the contralateral side of the body.

   a. The precentral gyrus contains a motor homunculus, which is a distorted orderly map of the skeletal muscles controlled by upper motor neurons in the contralateral limbs and trunk.

   b. In the motor homunculus, upper motor neurons that control cranial nerve-innervated muscles are represented close to the lateral fissure. Cranial nerve-innervated musculature and muscles of the hands are most heavily represented in the homunculus.

   c. Proceeding dorsally on the lateral aspect of the hemisphere are upper motor neurons that control neck, upper limb, and trunk musculature.

   d. On the medial aspect of the hemisphere is an extension of the primary motor cortex containing upper motor neurons that control pelvic and lower limb muscles.

   e. The primary motor cortex is known as M1; its upper motor neurons encode for the direction, force, and velocity of a movement and control individual muscles used in the execution of skilled movements.

LESIONS LIMITED TO M1

Lesions limited to M1 *are rare; these patients may have a contralateral weakness most evident in distal muscles that may or may not be accompanied by hyperreflexia and a Babinski sign (Figure 8–2B).*

2. The premotor cortex (area 6) is anterior to and more extensive than the primary motor cortex.

   a. Premotor cortical neurons are particularly active before neurons in the primary motor cortex and function in the planning and anticipation of complex motor acts.

   b. The basal ganglia and cerebellum provide much of the input to the premotor cortex.

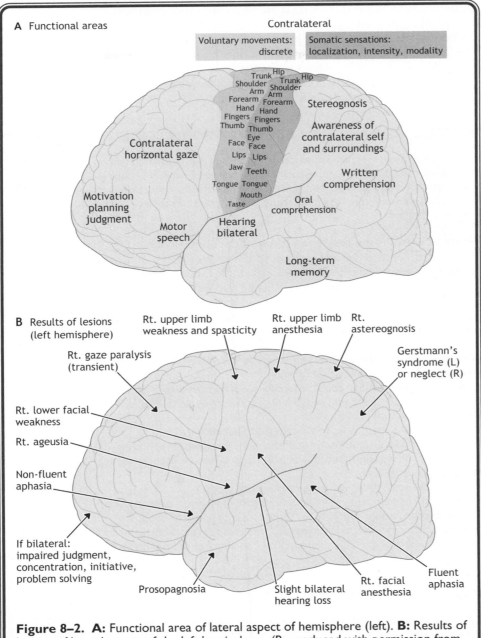

A Functional areas

Contralateral

Voluntary movements: discrete

Somatic sensations: localization, intensity, modality

Hip
Trunk Hip
Trunk
Shoulder Shoulder
Arm Arm
Forearm Forearm
Hand Hand
Fingers Fingers
Thumb Thumb
Eye
Face Face
Lips Lips
Jaw Teeth
Tongue Tongue
Mouth
Taste

Stereognosis

Awareness of contralateral self and surroundings

Contralateral horizontal gaze

Written comprehension

Motivation planning judgment

Oral comprehension

Motor speech

Hearing bilateral

Long-term memory

B Results of lesions (left hemisphere)

Rt. upper limb weakness and spasticity

Rt. upper limb anesthesia

Rt. astereognosis

Rt. gaze paralysis (transient)

Gerstmann's syndrome (L) or neglect (R)

Rt. lower facial weakness

Rt. ageusia

Non-fluent aphasia

If bilateral: impaired judgment, concentration, initiative, problem solving

Prosopagnosia

Slight bilateral hearing loss

Rt. facial anesthesia

Fluent aphasia

Figure 8–2. A: Functional area of lateral aspect of hemisphere (left). **B:** Results of lesions of lateral aspect of the left hemisphere. (Reproduced with permission from Young, PA. et al. *Basic Clinical Neuroanatomy*, LWW, 2007.)

3. The supplementary motor cortex (MII) is found on both the lateral and medial aspects of the frontal lobe.

    a. Upper motor neurons in the supplementary motor cortex function in mental rehearsal of movements before performing a complex motor task.

    b. Both supplementary motor cortex and premotor cortex function together in translating the desire to perform a motor task into a series of motor commands that accomplish the task.

DAMAGE TO PREMOTOR CORTEX AND SUPPLEMENTARY MOTOR CORTEX

Damage to premotor cortex and supplementary motor cortex may result in an *apraxia*, a disruption of the patterning and execution of learned motor movements. Individual movements are intact, and there is no weakness, but the patient is unable to perform learned movements in the correct sequence. Commonly, *lateral frontal lobe lesions* affect both primary motor and premotor cortical areas and result in weakness combined with apraxia.

    4. The **prefrontal cortex** is the association cortex of the frontal lobe.

        a. The prefrontal cortex is located in front of the premotor area and represents about 25% of the entire cerebral cortex in the human brain.

        b. The prefrontal cortex functions in organizing and planning the intellectual and emotional aspects of behavior.

        c. Parts of the prefrontal cortex are linked to limbic pathways and to the cognitive or executive basal ganglia circuit.

- *Lesions in the prefrontal area* result in *frontal lobe syndrome*. Patients may exhibit changes in behavior, intelligence, cognitive or executive functions, and memory.
- *Behavior changes* include a lack of appreciation and disregard for social rules, emotional withdrawal, a decrease in motivation, and abulia. Patients with abulia have a condition in which they lack sufficient levels of awareness to initiate a change in behavior.
  –Patients may test normally for intelligence but perform in an unintelligent manner.
- *A decrease in cognitive abilities* includes an inability to modify behavior in response to changing stimuli, impaired problem solving, and an inability to organize or plan. Patients may exhibit perseveration, an abnormal repetition of specific behaviors.
  –Patients with prefrontal cortex lesions may have short-term or working memory loss.
  –Premotor lesions may also result in the emergence of *infantile suckling or grasp reflexes* that are suppressed in adults. In the *suckling reflex*, touching the cheek causes the head to turn toward the side of the stimulus as the mouth searches for a nipple to suckle. In the *grasp reflex*, touching the palm of the hand results in a reflex closing of the fingers, which allows an infant to grasp anything that touches the hand.

CHANGES IN THE ACTIVITY OF NEURONS IN THE FRONTAL LOBE

- Changes in the activity of neurons in the frontal lobe may be the cause of *unipolar and bipolar disorders*, the 2 most common mood disorders.
- Unipolar depression is the most common mood disorder and may result from decreased neuronal activity in the frontal lobes inferior to the genu of the corpus callosum.
- Patients with bipolar disorder experience recurrent episodes of depression and euphoria or mania. In periods of euphoria, there is increased activity in the frontal lobes inferior to the genu of the corpus callosum.

5. The frontal eye field (Brodmann area 8) facilitates the generation of contralateral saccadic eye movements by activating neurons in brainstem gaze centers (Figure 8–2A).

6. Broca's area, in the left or dominant hemisphere, is the center for motor speech and corresponds to Brodmann areas 44 and 45 (Figure 8–2A).

    a. Neurons from Broca's area project to upper motor neurons in the adjacent primary motor cortex that control the vagus and hypoglossal nerves.

    b. The vagus nerves innervate muscles of the palate and larynx for speech. Tongue muscles innervated by the hypoglossal nerve are also used in speech.

LESIONS IN THE LATERAL ASPECT OF THE LEFT HEMISPHERE (FIGURE 8–2B)

Aphasias are disorders of language that result mainly from lesions in the lateral aspect of the left hemisphere. Aphasias may be caused by a stroke involving superficial branches of the left middle cerebral artery (MCA). Patients with an aphasia commonly have agraphia and difficulty in repetition or naming.

DAMAGE TO BROCA'S AREA

*Damage to Broca's area produces a **motor, nonfluent, or expressive aphasia**, which results in difficulty in putting together words to produce expressive speech.*

- *Patients with expressive aphasia can understand written and spoken language, but their verbal output is usually reduced to single-syllable words.*
- *Patients with expressive aphasia also have agraphia, although the hand used for writing can be used normally in other tasks.*
- *Patients with expressive aphasia are aware of and frustrated by their aphasia and have difficulty in repetition because of their lack of the ability to express their thoughts verbally or in writing.*

LESION OF BROCA'S AREA

*A **lesion of Broca's area** may include the adjacent primary motor cortex and result in a lesion of corticobulbar neurons arising from the left hemisphere and weakness of the muscles of the lower face on the right. If the lesion is large, there may be a spastic hemiparesis of the right upper limb. If the left frontal eye field is involved, the patient will look to the left, away from a paralyzed right upper limb.*

LESIONS IN BROCA'S AREA IN THE RIGHT HEMISPHERE

*Lesions in the area equivalent to Broca's (and Wernicke's area) in the right hemisphere are commonly caused by a **stroke involving superficial branches of the right middle artery** and result in **dysprosody**. These patients have normal speech and comprehension of speech but cannot express or do not comprehend the emotional and tonal qualities of speech that are crucial to verbal communication.*

    B. The **parietal lobe** functions in sensory processing and includes the primary and secondary somatosensory cortices, cortical areas for language comprehension, and the association cortex, which integrates somatosensory with other sensory modalities (see Figure 8–2A).

    1. The postcentral gyrus (Brodmann areas 3, 1, and 2) contains the primary somatosensory cortex (S1).

a. The S1 functions in cortical perception of discriminative touch, vibration, position sense, and pain and temperature sensations from the contralateral side of the face, scalp, neck, trunk, and limbs.
   b. The primary somatosensory cortex contains a sensory homunculus of the surface of the body that is distorted by the uneven cutaneous distribution of sensory receptors (see Figure 1–9).
   c. The fingertips, hands, and lips are the most heavily represented areas of the homunculus in which the density of sensory receptors is greater and the receptive fields are smaller than in other parts of the body.
   d. The sensory homunculus has a somatotopic representation similar to the homunculus in primary motor cortex.
      (1) The head, neck, upper limb, and trunk are represented on the lateral aspect of the hemisphere.
      (2) The pelvis and lower limb are represented in an extension of the post-central gyrus onto the medial aspect of the hemisphere (Figure 8–3A).
   2. A separate topographic map exists in each of 3 somatosensory cortical Brodmann areas—3, 1, and 2—in which each cortical neuron is defined by its receptive field and its sensory modality.
      a. Area 3 receives most of the projections from the thalamus; areas 1 and 2 represent cortical areas for higher-order somatosensory processing.
      b. Neurons in area 3 respond to muscle stretch input provided by muscle spindle afferents and to rapidly adapting and slowly adapting cutaneous inputs, which are important for discrimination of the size, shape, and texture of a stimulus.
      c. Neurons in a subset of area 3 provide information as to which phalanx of which finger is in contact with an object.
      d. Neurons in area 1 respond to the texture of a stimulus as detected by multiple fingers.
      e. Neurons in area 2 respond to complex features of size and shape such as the curvature of the object, the spacing of ridges on textured surfaces, or the direction of motion of the object stimulus across the hand.

LESIONS IN THE SOMATOSENSORY CORTEX

*Lesions in the somatosensory cortex result in **impairment in the perception of somatic sensations** on the opposite side of the face, scalp, trunk, and limbs. The primary somatosensory cortex may reorganize after injury to peripheral receptors or to the primary sensory neurons that innervate them. For example, the loss of sensory input from an amputated digit 3 results in the cortical area responsive to that digit reorganizing so that its neurons now respond to the adjacent digits 2 and 4.*

   3. A second somatosensory cortical area called S2 contains 2 additional somatosensory maps and is situated on the superior bank of the lateral fissure just below S1.
      a. The S2 cortex also receives thalamic input but is driven by neurons in all 3 regions of S1 and by neurons from the somatosensory cortex in the opposite hemisphere by axons that traverse the corpus callosum.
      b. The S2 cortex is a site in which somatosensory information is processed from both sides of the body. S2 is involved in permitting a tactile discriminatory task that is learned by one hand to be performed by the other.

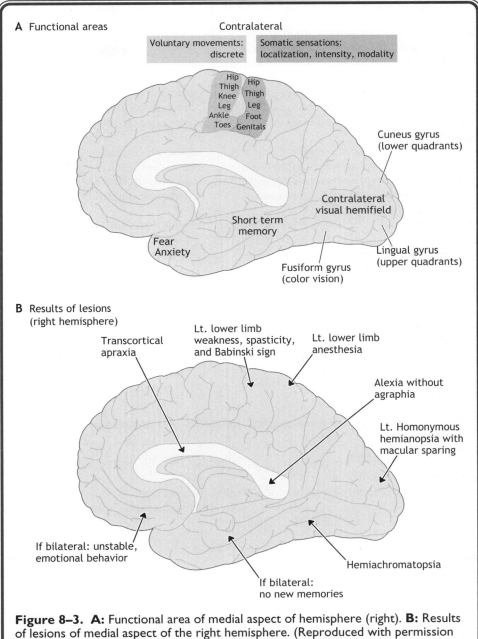

Figure 8–3. A: Functional area of medial aspect of hemisphere (right). **B:** Results of lesions of medial aspect of the right hemisphere. (Reproduced with permission from Young PA, et al. *Basic Clinical Neuroanatomy,* LWW, 2007.)

c. Neurons in S2 project to the insular cortex, which in turn projects to parts of the limbic system involved in the memory of the tactile stimulus.

   4. Sensations of pain and temperature are perceived in the primary somatosensory cortex, insular cortex, and cingulate gyrus.

    a. The insular cortex integrates pain with other sensory inputs and projects the pain information into the limbic system.

    b. The cingulate gyrus processes an emotional content of pain.

   5. The superior parietal lobule (Brodmann areas 5 and 7) is the parietal association cortex that integrates somatosensory input with other sensory modalities and assists in motor planning.

    a. Area 7 integrates visual, somatosensory, and auditory modalities for motor activities that require hand–eye coordination.

    b. Area 5 receives information from joints and from groups of muscles that act at those joints to provide information as to where the contralateral limbs are in space.

    c. Premotor cortex uses information from areas 5 and 7 in motor planning.

LESIONS TO THE SUPERIOR PARIETAL LOBULE

- *Lesions of the superior parietal lobule association areas, including areas 5 and 7, usually in the dominant hemisphere, may result in one of several forms of **apraxia or astereognosis** (see Figure 8–2A and B).*
- *These patients may have an **ideational apraxia**, a lack of understanding of how to organize the sequence of a pattern of movements.*
- *These patients may have an **ideomotor apraxia**, in which they cannot perform tasks on command, even though there is no motor weakness. They may be able to identify an object correctly but will not know how to use it.*
- *The patient may have a **constructional apraxia**, in which they are unable to copy a simple diagram or describe how to get from their home to the store.*
- *The patient may have **astereognosia**, an inability to recognize an object held by its size and shape without looking at the object.*

   6. The inferior parietal lobule contains the angular gyrus (area 39) and the supramarginal gyrus (area 40).

    a. The angular gyrus (area 39) is a center for written comprehension.

    b. Wernicke's area (Brodmann area 22) in the adjacent temporal lobe is an oral comprehension center (Figure 8–2A and B)

LESIONS IN WERNICKE'S AREA

*Lesions in Wernicke's area (area 22) of the temporal lobe result in a fluent, receptive, sensory, or **Wernicke's aphasia** (see later discussion).*

LESION TO THE ANGULAR GYRUS

*A lesion to the angular gyrus (area 39) in the left inferior parietal lobule may result in a loss of ability to comprehend written language (**alexia**) and to write it (**agraphia**), but spoken language may be understood.*

LESION TO THE INFERIOR PARIETAL LOBULE

*Patients with a lesion to the inferior parietal lobule, including the angular gyrus in the left hemisphere, may have **Gerstmann's syndrome**. These patients have **acalculia** (an inability to perform simple arithmetic problems), **finger agnosia** (an inability to recognize one's fingers), and **right-left disorientation**. Alexia with agraphia may also be present.*

LESION IN AREAS 39 AND 40

*A lesion in the inferior parietal lobule of the right hemisphere results in **contralateral neglect**.*

- *These patients lack awareness of or neglect the contralateral half of the body (**asomatognosia**). Although somatic sensation is intact, the patients ignore the left half of their body and may fail to dress, undress, or take care of the affected side.*
- *If these patients are asked to draw the numbers on a clock face from memory, they will draw all 12 numbers on the right side, ignoring the left half of the clock face. These patients may deny that an arm or leg belongs to them when the affected limb is brought into their field of vision or may deny their deficit entirely (**anosognosia**).*
- ***Unilateral neglect** is uncommon in patients with a left parietal lobe lesion because of compensation of perceptual awareness by the right hemisphere.*

    C. The **occipital lobe is essential for the reception and recognition of visual stimuli** and contains primary visual and visual association cortex (see Figures 8–1B and 8–3A and B).

        1. The visual cortex is divided into striate or primary visual (V1 or area 17) and extrastriate (areas 18 and 19) cortices.

        2. Area 17, the primary visual cortex, lies on the medial portion of the occipital lobe on either side of the calcarine sulcus.

            a. The retinal surface (and, therefore, the visual fields) is represented in a topographic manner in area 17.

            b. The peripheral parts of the contralateral nasal and temporal hemifields are represented in the anterior part of primary visual cortex.

            c. The central parts of each visual field representing the area of the macula are in the posterior part of primary visual cortex near the tips of the occipital lobes.

CORTEX LESIONS AND VISUAL FIELD DEFICITS

*A complete unilateral lesion of primary visual cortex results in a **contralateral homonymous hemianopsia with macular sparing**, usually resulting from an infarct of a branch of the posterior cerebral artery (PCA). The area of the macula of the retina containing the fovea is spared because of a dual blood supply from the PCA and MCA. Bilateral visual cortex lesions result in cortical blindness (Anton syndrome).*

INJURY TO THE BACK OF THE HEAD

*A blow to the back of the head may result in a **loss of macular representation of the visual fields**.*

        3. Extensive areas of visual association cortex are rostral to the primary visual cortex and extend into the posterior parts of the parietal and temporal lobes.

4. Form and color are processed in the occipital and temporal lobes by two parvocellular systems.
   a. The parvocellular streams relay through parvocellular layers of the lateral geniculate and project to blob/interblob zones of primary visual cortex.
   b. These areas project to the inferior part of the temporal lobe to Brodmann areas 37 (fusiform gyrus), 20 and 21, and form the "what" visual pathway.

UNILATERAL LESIONS IN THE LINGUAL (FUSIFORM) GYRUS (FIGURE 8–3B)

*Unilateral lesions in the lingual gyrus (area 37) result in **hemiachromatopsia**, in which the patient notes a change in the appearance of colors. Bilateral lesions result in a **complete loss of color vision** so that the patient sees only in shades of gray.*

5. Motion and depth are processed in the parietal lobe by a magnocellular system.
   a. The magnocellular stream relays through magnocellular layers of the lateral geniculate and projects to non-blob zones of primary visual cortex.
   b. These areas project into the parietal lobe through Brodmann area 18 to area 7 and form the "where" visual pathway.

PARIETAL LESIONS AFFECTING MOTION AND DEPTH

- *Parietal lesions may result in **Balint syndrome**, which includes visual disorientation, ocular apraxia, and optic ataxia.*
- *Patients with **visual disorientation** cannot appreciate more than one aspect of a visual scene at a time.*
- *Patients with **ocular apraxia** cannot focus on an object of interest, have difficulty initiating saccades, and tend to overshoot or undershoot visual targets.*
- *Patients with **optic ataxia** have difficulty reaching for an object under visual control but are able to touch their nose easily with a finger.*

D. The temporal lobe contains primary auditory cortex, Wernicke's area, limbic structures (see Chapter 6), limbic cortex, and association cortex (see Figures 8–2A and B and 8–3A and B).
   1. The primary auditory cortex (Brodmann areas 41 and 42) is located on the transverse **gyri of Heschl,** in the superior temporal lobule deep within the lateral fissure.
   2. Wernicke's area (area 22) is an oral comprehension center that is directly posterior to the primary auditory cortex.

UNILATERAL DAMAGE TO THE PRIMARY AUDITORY CORTEX

*Patients with unilateral damage to the primary auditory cortex may have a slight **bilateral hearing loss** and **difficulty localizing the source of a sound**.*

LESIONS IN AREA 22 IN THE TEMPORAL LOBE (FIGURE 8–2B)

*Lesions in area 22 in the temporal lobe result in a fluent, receptive, or **Wernicke's aphasia**.*

- *Patients with Wernicke's aphasia cannot comprehend spoken language and may or may not be able to read (alexia), depending on the extent of the lesion. The deficit is characterized by fluent speech, but it lacks meaning. Patients may speak in complete sentences but frequently misuse words.*

- *Patients with a Wernicke's aphasia have difficulty with repetition because they do not understand the command, are generally unaware of their deficit, and are not concerned about their condition.*
- *In deaf patients who communicate by sign language, a lesion to Broca's or Wernicke's area will result in a corresponding motor or sensory sign aphasia.*

3. The arcuate fasciculus connects Broca's area in the frontal lobe with Wernicke's area in the temporal lobe (see Figure 8–3A).

LESION OF THE ARCUATE FASCICULUS

*A lesion of the arcuate fasciculus results in a **conduction aphasia**, a disconnect syndrome in the left hemisphere. Verbal output is fluent, but the patient misuses words. Language comprehension is normal, but the patient cannot repeat words or execute verbal commands by an examiner (such as counting backward beginning at 100) and demonstrates poor object naming. Like an expressive aphasia, the patient is aware of the deficit and frustrated by the inability to execute a verbal command that was heard and understood.*

4. Temporal association cortex functions in recognition, in particular, by using the cone visual stream to identify and name objects.

LESIONS IN THE TEMPORAL ASSOCIATION CORTEX

- *Lesions in the parvocellular stream to temporal association cortex result in **visual agnosias**.*
- *These patients acknowledge the existence of an object that they feel or see (unlike patients with neglect) but cannot identify the object using visual input (object agnosia).*
- *Patients with lesions of the temporal association cortex may develop **prosopagnosia**, an inability to recognize familiar people by sight. As soon as the person speaks, however, he or she is recognized.*

VI. Commissural fibers

A. Commissural fibers mainly consist of axons that interconnect similar regions in the opposite hemisphere.

B. The **corpus callosum, anterior commissure, and posterior commissure** contain the commissural fibers.

1. The corpus callosum is a massive commissural system that interconnects corresponding areas of the frontal, parietal, temporal, and occipital lobes.

a. The corpus callosum consists of a rostrum, genu, body, and splenium (Figure 8–1B).

b. The rostrum, genu, and body are supplied by the ACA; the splenium is supplied by the PCA.

2. The anterior commissure contains axons of the medial olfactory tracts, which interconnect the olfactory bulbs.

3. The posterior commissure contains axons of the pretectal nuclei, which bilaterally innervate preganglionic neurons in the Edinger-Westphal nucleus in the midbrain. The posterior commissure is commonly compressed in patients with Parinaud's syndrome (see Chapter 6).

CORPUS CALLOSUM LESIONS

*Lesions of commissural fibers in parts of the corpus callosum result in **disconnect syndromes** in which regions of the 2 hemispheres can no longer communicate with each other.*

LESIONS TO THE BODY OF THE CORPUS CALLOSUM

- Lesions to the body of the corpus callosum may result in a **transcortical ideomotor apraxia** caused by an occlusion of the ACA. As in other forms of apraxia, there is no motor weakness but the patient cannot execute a command to move the left arm. The patient understands the command, which is perceived in Wernicke's area of the left hemisphere, but the callosal lesion disconnects Wernicke's area from the right motor cortex, so that the command cannot be executed. The patient is still able to execute a command to move the right arm because Wernicke's area in the left hemisphere is able to communicate with the left motor cortex without using the corpus callosum.

LESION TO THE SPLENIUM (FIGURE 8–3B)

- A lesion to the splenium of the corpus callosum caused by an occlusion of the left PCA may result in **alexia without agraphia**. Alexia without agraphia is a disconnect syndrome that prevents visual information in the right occipital cortex from reaching language comprehension centers in the left hemisphere. Patients can see words in the left visual field but do not understand their meaning.
- These patients are unable to read and often have a color anomia (inability to name colors). These patients can write, but they cannot read what they wrote. A lesion of the left occipital cortex may also result in a **right homonymous hemianopsia with macular sparing**.
- **Alexia with agraphia**, an inability to read or write, may result from a lesion of the angular gyrus in the left parietal lobe.

VII. The internal capsule

- A. The **internal capsule** contains axons of virtually all neurons that enter or exit the cortex.
- B. The internal capsule may be divided into an **anterior limb, a posterior limb, and a genu** (Figures 8–4 and 8–5).
  - 1. The anterior limb is situated between the head of the caudate and the putamen and globus pallidus of the basal ganglia.
    - a. The anterior limb contains descending axons from the prefrontal cortex and thalamocortical axons from the limbic thalamic nuclei.
    - b. Medial striate branches of the anterior cerebral and lenticulostriate branches of the MCA supply the anterior limb of the internal capsule.
  - 2. The genu of the internal capsule is at the level of the interventricular foramen of Monro and contains corticobulbar axons.
  - 3. The posterior limb is situated between the basal ganglia and the thalamus.
    - a. The posterior limb contains corticospinal axons with a topographic organization (see Figure 8–5); arm (A), then leg (L).
    - b. The posterior limb contains thalamocortical somatosensory axons from the ventrobasal complex that are caudal to the corticospinal axons and have a similar topographic arrangement (face, arm, leg).
    - c. The posterior limb also contains auditory projections from the medial geniculate nucleus and the proximal parts of the visual radiations from the lateral geniculate nucleus.
    - d. The anterior choroidal artery supplies the genu and posterior limb of the internal capsule.

INTERNAL CAPSULE LESIONS

A **lacunar stroke** involving the posterior limb of the internal capsule may result in a complete contralateral anesthesia, contralateral hemiplegia, contralateral homonymous hemianopsia, and a slight bilateral hearing loss.

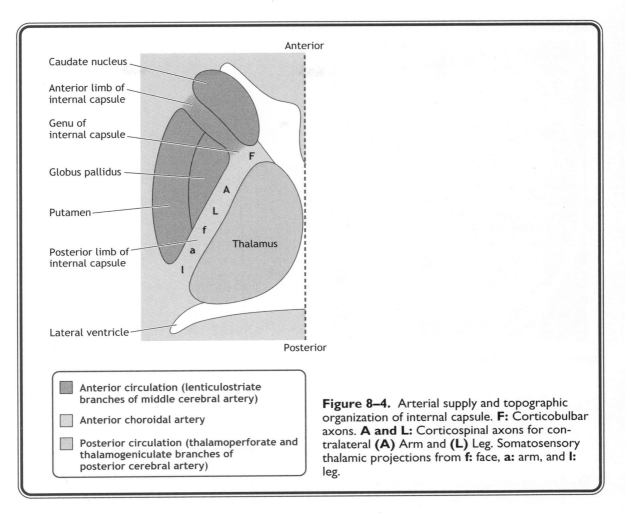

Figure 8–4. Arterial supply and topographic organization of internal capsule. **F:** Corticobulbar axons. **A and L:** Corticospinal axons for contralateral **(A)** Arm and **(L)** Leg. Somatosensory thalamic projections from **f:** face, **a:** arm, and **l:** leg.

A lacunar stroke involving the genu of the internal capsule may result in contralateral lower face weakness, transient deviation of the uvula toward the lesion, and deviation of the tongue away from the lesion on protrusion.

VIII. Electroencephalogram to measure cortical activity

    A. An **electroencephalogram** (EEG) provides a crude assessment of cortical activity to measure the postsynaptic potentials of cortical neurons.

    B. An EEG is measured by attaching 19 electrodes to the scalp and 1 to each auricle; **amplifiers detect the differences in electrical activity between pairs of electrodes.**

    C. An EEG is restricted to a narrow frequency range of 0–60 Hz divided into 4 bands.

        1. The **alpha rhythm** is 8–13 Hz and is recorded in awake individuals.

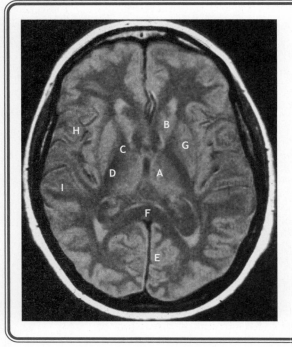

Figure 8–5. MRI of horizontal section through diencephalon, basal ganglia, and cortex. **A:** Thalamus. **B:** Head of caudate nucleus. **C:** Genu of internal capsule containing corticobulbar axons. **D:** Posterior limb of internal capsule (see Figure 8–4). **E:** Primary visual cortex. **F:** Splenium of corpus callosum. **G:** Putamen. **H:** Broca's motor speech area. **I:** Wernicke's oral comprehension area.

2. **Beta activity** is from 14–60 Hz and indicates increased mental activity and attention.
3. **Theta waves** (4–7 Hz) and **delta waves** (< 4 Hz) are prominent during non–rapid eye movement (non-REM) sleep, drowsiness, or pathological conditions.

IX. The ascending arousal system

A. The **ascending arousal system** increases arousal and facilitates consciousness by maintaining a desynchronized EEG.

1. **Arousal** is the level of wakefulness determined by the response levels of thalamic and cortical neurons to sensory stimuli.
2. **Consciousness** is the level of one's own awareness and place in the environment.

B. The ascending arousal system consists of monoaminergic neurons (noradrenergic, serotonergic, histaminergic neurons) and cholinergic neurons in the brainstem and hypothalamus that project to the thalamus and cortex.

1. Important neurons in the maintenance of arousal are cholinergic neurons in the pedunculopontine nucleus at the midbrain pons border and in the basal nucleus of Meynert.
2. These cholinergic neurons project to the diencephalon and cortex.

C. The ascending arousal system has 2 components: one projects through the thalamus to the cerebral cortex, and the other projects through the hypothalamus to the cerebral cortex.

LESION OF EITHER COMPONENT OF THE ASCENDING AROUSAL SYSTEM

- *A lesion of either component of the ascending arousal system may **disrupt consciousness** and **may result from lesions of the midbrain, diencephalon, or cortex.** Lesions caudal to the pons generally do not affect consciousness.*
- *Patients are in a **coma** if they do not make any response to a strong sensory stimulus (e.g., rubbing the skin over the sternum with a knuckle). The EEG of a comatose patient exhibits fixed patterns that do not vary cyclically like the EEG during non-REM and REM sleep (see following discussion).*

X. Sleep

A. Sleep consists of **non-REM stages** that vary cyclically with an **REM stage** during which the EEG cycles between a desynchronized state and a synchronized state.

B. During sleep, monoaminergic and cholinergic neurons in the ascending arousal system have different levels of activity.

C. During a sleep cycle, one passes through non-REM sleep and then through REM sleep (Figure 8–6).

  1. There are usually 4–6 sleep non-REM/REM cycles in an 8–h period of sleep.

  2. Seventy-five to 80% of a sleep cycle is spent in non-REM sleep; the amount of time spent in non-REM sleep decreases with each cycle.

  3. Twenty to 25% of a sleep cycle is spent in REM sleep; the amount of time spent in REM sleep increases with each cycle.

D. Non-REM sleep is divided into 4 stages (I–IV) characterized by a progressively synchronized EEG combined with awake-state levels of muscle tone that result in an inactive nervous system in an active body.

  1. Non-REM sleep is induced by "non-REM-on" gamma-aminobutyric acid (GABA) neurons in the preoptic area of the hypothalamus and by serotonin from the raphe nuclei.

  2. During non-REM sleep, brainstem noradrenergic neurons show decreased activity.

  3. Stage I is a period of drowsiness that lasts 10 min or less in which the EEG is dominated by 4- to 7-Hz theta waves. Individuals are easily aroused from stage I non-REM sleep.

  4. Stage II is characterized by sleep spindle complexes, 10- to 15-Hz oscillations in the EEG that arise from periodic interactions between thalamic and cortical neurons.

  5. Stages III and IV are known as slow-wave, delta, or deep sleep, in which the synchronized EEG shows prominent 1- to 4-Hz delta waves; it is difficult to awaken people from stage IV sleep.

  6. During non-REM sleep, skeletal muscles relax but maintain their tone. The parasympathetic nervous system is active and promotes gastric motility and a decrease in heart rate and blood pressure.

LESIONS OF THE POSTERIOR HYPOTHALAMUS

Lesions of the posterior hypothalamus or antihistamines that block the activity of histaminergic neurons in the posterior hypothalamus promote sleepiness.

E. REM, or paradoxical, sleep is characterized by a sudden conversion from a synchronized to a desynchronized EEG, resembling that of the awake state

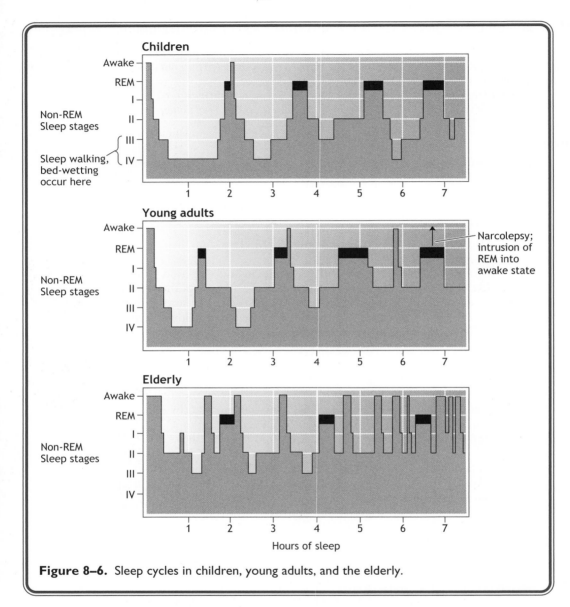

Figure 8–6. Sleep cycles in children, young adults, and the elderly.

combined with a loss of muscle tone, which results in an active nervous system in an inactive body.
1. REM sleep commences only after non-REM sleep progresses through stages I–IV and back to stage II, a cycle that lasts about 90 min.
2. Brainstem cholinergic neurons in the ascending arousal system act as "REM-on" cells that initiate REM sleep.

3. During REM sleep, inhibitory medullary reticulospinal neurons are activated, causing skeletal muscles to become flaccid and muscle stretch reflexes to be absent; only the ocular, respiration, and middle ear muscles remain active.
4. During REM sleep, all sensory systems are inhibited.
5. During REM sleep, there is an increase in blood pressure, metabolism, and blood flow to the brain.
6. Penile and clitoral erections are features of REM sleep.
7. REM sleep is associated with rapid saccadic conjugate eye movements and by PGO waves in the EEG generated by **pontine cholinergic projections** to the lateral **geniculate nucleus** of the thalamus and to the **occipital cortex**.
8. Most dreams occur during REM sleep, and the visual images perceived during dreaming are associated with PGO waves in the EEG.
9. Near the end of an REM sleep period, noradrenergic and serotonergic neurons act as "REM-off" cells, initiating a transition from REM sleep back to non-REM sleep.

CHANGES IN SLEEP PATTERNS WITH AGING

With aging, there is a **decline in the length of time spent in sleep** and a **decrease in both time spent in non-REM and in REM sleep**. The elderly spend little time in stage IV slow-wave sleep and exhibit frequent brief awakenings after periods of REM sleep.

SLEEP DISORDERS

- **Insomnia** is an inability to initiate or obtain enough sleep and may be a symptom of anxiety, depression, chronic pain, or drug abuse. The EEG of an insomniac shows no irregularities, but these patients do not seem to benefit from a period of sleep.
- **Sleep apnea** occurs when breathing stops for up to 30 seconds during sleep and is associated with loud snoring. Obstructive sleep apnea may be caused by abnormalities that narrow the respiratory pathway, such as a deviated nasal septum or enlarged tonsils. Central sleep apnea results from an inhibition of brainstem respiratory centers. Polysomnograms are used to distinguish whether a patient has a central or an obstructive sleep apnea.
- **Narcolepsy** results from an intrusion of REM sleep into the awake state. Patients with narcolepsy have daytime sleep attacks, cataplexy, a sudden loss of muscle tone and control (which may cause them to fall if standing), dreamlike hallucinations, and sleep paralysis (after awaking, the patient is unable to move for several minutes).
- Sleepwalking, sleep terrors, and nocturnal bed-wetting are **parasomnias** that occur during non-REM stage III or stage IV sleep.

XI. Seizures

A. **Seizures** result from cortical neurons that transiently fire in an abnormal, synchronized, high-frequency manner and may be partial or generalized.

B. **Partial (local or focal) seizures** are localized to a part of the brain. The symptoms experienced depend on the affected area of the brain.

C. **Simple partial seizures** typically last 5–10 seconds (the ictal period), consciousness is preserved, and there are usually no postictal deficits.

1. Patients may have **motor symptoms** ranging from a twitching of a contralateral hand to jerky clonic movements or sustained tonic contraction of the contralateral limbs.

2. Patients with **visual cortex seizures** may see flashes of light or have visual hallucinations.

3. Patients with **auditory cortex seizures** may hear voices, music, or roaring sounds.

4. Patients with **somatosensory cortex seizures** may have paresthesias in the contralateral limbs.

D. **Complex partial seizures** typically last from 30 seconds to 2 minutes, consciousness is impaired, and there may be postictal deficits.

   1. The most common location for a complex partial seizure is in the temporal lobe.

   2. An **aura is a simple partial seizure**, which may precede a complex partial seizure. During the simple partial seizure, the patient may have olfactory, emotional, visceral, or déjà vu sensations.

   3. Patients may exhibit a level of unresponsiveness and lack of awareness.

   4. Patients may have **automatisms**, during which they exhibit repetitive behaviors such as smacking of the lips, swallowing, or stroking or patting movements.

   5. **Postictal deficits** may include amnesia, aggression, depression, or headache.

E. **Generalized seizures** result from abnormal firing of neurons in many areas of the CNS.

   1. The **most common generalized seizure is a grand mal, Jacksonian, or tonic-clonic seizure**, which begins in the primary motor or premotor cortex on one side and then spreads across the corpus callosum to the opposite hemisphere.

     a. During a **tonic phase**, which lasts 10–15 seconds, there are contractions of the fingers of one hand.

     b. The tonic phase is followed by a **clonic phase**, lasting 30 seconds to 2 minutes, during which there are contractions of skeletal muscles in the arm, face, and lower limb and then rhythmic, jerky bilateral contractions of both limbs. These contractions coincide with a loss of consciousness as the seizure spreads to the ascending arousal system.

     c. There may be autonomic effects, including excessive salivation, dilated pupils, hypertension, and urinary incontinence.

     d. In an initial postictal period, patients are immobile and unresponsive, lie with their eyes closed, and breathe deeply.

   2. An **absence or petit mal seizure** is a generalized seizure with negative symptoms characterized by a brief period of unresponsiveness and a lack of knowledge of what happened during the seizure.

     a. Absence seizures are not preceded by an aura.

     b. Absence seizures are more common in children than in adults.

EPILEPSY

Epilepsy is a central nervous system (CNS) disorder characterized by recurrent repetitive seizures. A patient who has status epilepticus has continuous episodes of rapidly repetitive seizures that last longer than 30 minutes.

XII. Arterial blood supply of the cerebral cortex

A. The **MCA** passes into the lateral fissure and supplies most of the lateral convexity of each hemisphere.

B. In the lateral fissure, the MCA divides into **2 superficial branches**: a superior division and an inferior division.

    1. The **superior division** supplies the lateral aspect of the frontal lobe and most of the precentral and postcentral gyri on either side of the central sulcus. The superior division of the left MCA also supplies Broca's area, the motor speech center.

    2. The **inferior division** supplies the superior and anterior parts of the temporal lobe and part of the parietal lobe. The left inferior division of the MCA supplies Wernicke's area and the angular gyrus, which are language comprehension centers.

DEFICITS ASSOCIATED WITH AN OCCLUSION OF THE MCA

- An **occlusion of either MCA** may result in contralateral upper limb and face hemiparesis, contralateral upper limb and face hemianesthesia, contralateral homonymous hemianopsia, and deviation of the eyes toward the affected hemisphere.
- Patients with left MCA involvement may also have a Broca's, Wernicke's, conduction, or global aphasia. Those with right MCA involvement may have contralateral hemineglect or dysprosody.

C. The **anterior cerebral artery** (ACA) is smaller than the MCA and **courses on the medial aspect of the hemisphere through the interhemispheric fissure**.

    1. The superficial branches of the anterior cerebral supply the medial aspects of the frontal lobe adjacent to the olfactory bulb and tract.

    2. The superficial branches supply the parietal lobe, including medial extensions of the precentral and postcentral gyri.

DEFICITS ASSOCIATED WITH AN OCCLUSION OF THE ACA

A **stroke** involving either ACA may result in contralateral lower limb hemiparesis, contralateral lower limb hemianesthesia, transcortical apraxia (if just the corpus callosum is involved), and frontal lobe behavioral deficits.

D. The **PCAs arise at the bifurcation of the basilar artery and course posteriorly to supply the posterior and inferior aspects of each hemisphere**.

E. The superficial branches of the PCA supply the inferior and medial parts of the lateral aspect of the temporal lobe and the lateral and medial aspects of the occipital lobe.

DEFICITS ASSOCIATED WITH AN OCCLUSION OF THE PCA

An **occlusion of either PCA** may result in **contralateral homonymous hemianopsia with macular sparing**. With involvement of the left PCA, which also supplies the splenium of the corpus callosum, patients may also have **alexia without agraphia**.

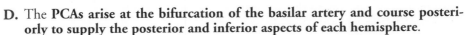

CLINICAL PROBLEMS

1. Your patient has suffered a stroke and seems to be suddenly unable to understand what is said to him. His speech seems normal, but does not make sense. What does the patient have?

   A. Gerstmann's syndrome

   B. Conduction aphasia

   C. Unilateral neglect

   D. Alexia without agraphia

   E. Receptive aphasia

2. A healthy 32-year-old woman suddenly experiences prolonged contractions of the finger muscles of the left hand that progress up her left arm to her face and down to her lower limb. She loses consciousness and falls to the floor. While unconscious, she exhibits jerky movements of all 4 limbs, bloody saliva drips from her mouth, and she loses bladder control. She slowly regains consciousness after 30 seconds, but her muscles remain flaccid. An hour later, she seems to have completely recovered. How would you characterize the patient's apparent seizure?

   A. Simple partial seizure

   B. Jacksonian seizure

   C. Absence seizure

   D. Petit mal seizure

   E. Complex partial seizure

3. Your patient has a language problem, the right corner of the mouth droops, and the right upper limb is weak. There are no sensory deficits. Which of the following will be evident in this patient?

   A. The patient may be unaware of the deficit.

   B. The patient can speak in complete sentences but misuses words.

   C. The patient may also have finger agnosia.

   D. The patient may have impaired repetition.

   E. The patient may also have a homonymous hemianopsia.

4. Your patient has suffered a stroke caused by occlusion of the superficial branches of the right ACA. The patient may have:

   A. A loss of pain and temperature sensations in the left upper limb

   B. A spastic weakness of the left lower limb

   C. Lower face weakness on the left

   D. Inability to look to the left with both eyes

   E. A loss of discriminative touch from the right side of the face

5. Your female patient has suffered a stroke. She has difficulty copying simple diagrams, even though she hears and understands your requests to do so. You notice that she only has makeup on the right side of her face and does not seem to know where her left hand is in space. Where is her lesion?

    A. Left prefrontal cortex
    B. Right parietal lobe
    C. Left temporal lobe
    D. Right occipital lobe
    E. Left parietal lobe

6. A 50-year-old man suddenly has difficulty speaking and develops weakness in the right upper limb. His speech is nonfluent, lacks melody, and can be produced only with great effort. Speech comprehension appears to be normal, but the man has difficulty writing. What else might you expect to observe in the patient?

    A. Anterograde amnesia
    B. Lower face weakness on the right
    C. Lack of patient awareness of the deficit
    D. A right superior quadrantanopsia
    E. Dysmetria

7. Your 18-year-old patient is participating in a sleep study and has entered REM sleep. Which of the following is true of the patient?

    A. The EEG has become desynchronized.
    B. Cholinergic neurons have decreased their activity.
    C. Sleepwalking and sleep terrors may occur.
    D. Seventy-five percent of the average period of sleep will be spent in REM sleep.
    E. The length of time spent in REM sleep will decrease as the sleep period progresses.

8. A man was brought to the hospital after collapsing. After he regained consciousness, an exam revealed that epicritic and protopathic sensations were lost on the left side of the body and on the left side of the face and scalp. The patient has a left spastic hemiplegia with increased deep tendon reflexes and Babinski's sign and weakness of the lower face on the left. The patient could communicate reasonably well and could understand commands. Which cause or site of a single lesion accounts for all of the symptoms?

    A. Stroke involving the left MCA, including the superior and inferior divisions
    B. Lacunar stroke involving the lenticulostriate branches of the MCA that supply the genu and posterior limb of the internal capsule on the right
    C. Watershed infarct between the vascular territories of the MCA and ACA
    D. Jacksonian seizure
    E. Neoplasm in the interhemispheric sulcus that compresses the ACAs

9. A 15-year-old boy was thrown from his bicycle and suffered head trauma. Computed tomographic (CT) imaging revealed no intracranial bleeding, and the patient was released from the hospital after 3 days. Two years later, the patient complained of brief periods when he perceived the smell of rotten eggs while seated at his desk at school. One morning the patient's high school teacher found him in the restroom shaking, with saliva dripping from his mouth. The patient's pupils were dilated, and the patient was sweating. After the incident, the patient seemed confused and drowsy and did not remember going to the restroom. How might a neurologist characterize the patient's neurological incident?

A. Absence seizure

B. Status epilepticus

C. Complex partial seizure

D. Simple partial seizure

E. Petit mal seizure

10. A middle-aged man suddenly has difficulty reading but hears and understands what is being said to him. He also has difficulty writing, even though the strength in his upper limbs is 5/5, and he is unable to pick up a pen using the thumb and index finger of his right hand. The patient seems to have lost the ability to add and subtract, despite being able to understand the request to do so. A diagnosis of the patient's condition might be:

A. Alzheimer's disease

B. Gerstmann's syndrome

C. Balint's syndrome

D. Frontal lobe syndrome

E. Transcortical apraxia

MATCHING PROBLEMS

Questions 11–27: Cortical lesions/features match

Choices: (each choice may be used once, more than once, or not at all):

A. Primary motor cortex on medial aspect of hemisphere

B. Precentral gyrus on lateral aspect of hemisphere

C. Area 44, 45 of dominant hemisphere

D. Angular gyrus area 39

E. Arcuate fasciculus

F. Wernicke's area 22

G. Splenium of corpus callosum

H. Brodmann areas 5 and 7

I. Brodmann area 8

J. Inferior parietal lobule of right hemisphere

K. Postcentral gyrus on lateral aspect of hemisphere

    L. Brodmann areas 41 and 42

    M. Cuneus gyrus

    N. Prefrontal cortex

    O. Body of corpus callosum

    P. Brodmann area 17

    Q. Primary somatosensory cortex on medial aspect of hemisphere

    R. "Wernicke's" area in right or nondominant hemisphere

11. Lesions here result in alexia without agraphia.

12. Lesions here might affect the ability to empty the bladder.

13. Lesions here result in contralateral homonymous hemianopsia with macular sparing.

14. Lesions here result in astereognosis.

15. Lesions here result in reduced verbal output and agraphia, but no weakness in limbs.

16. Lesions here result in an expressive aphasia.

17. Lesions here result in constructional apraxia.

18. Lesions here result in alexia without agraphia, finger agnosia, and dyscalculia.

19. Lesions here result in a conduction aphasia.

20. Lesions here result in anesthesia in contralateral lower limb.

21. Lesions here result in an inability to move the left arm in response to a command, but no motor weakness.

22. Lesions here result in lack of initiative, apathy, and other personality changes.

23. Lesions here result in lack of understanding of emotional content of speech.

24. Lesions here might result in a drooping of the corner of the mouth.

25. Lesions here result in an inability to look horizontally away from the side of the lesion.

26. Lesions here result in no visual field deficits, but patients fail to recognize their left hand as being theirs.

27. Lesions here result in Gerstmann's syndrome.

ANSWERS

1. The answer is E. Patients with Wernicke's aphasia cannot comprehend spoken language and may or may not be able to read (alexia). The deficit is characterized by flu-

ent speech that lacks meaning. Patients may speak in complete sentences, frequently misuse words, and have difficulty with repetition because they do not understand the command.

2. **The answer is B.** The patient had a grand mal, Jacksonian, or tonic-clonic seizure, which begins in the primary motor or premotor cortex on one side. It begins with contractions of the fingers of one hand followed by contractions of skeletal muscles in the arm, face, and lower limb and then rhythmic, jerky bilateral contractions of both limbs, which coincide with a loss of consciousness.

3. **The answer is D.** Patients with an expressive aphasia can understand written and spoken language, but their verbal output is usually reduced to single-syllable words. Patients with an expressive aphasia are aware of and frustrated by their aphasia and have difficulty in repetition because of an inability to express thoughts orally or in writing.

4. **The answer is B.** The ACA supplies somatosensory and motor cortical areas that represent the contralateral lower limb. All other choices are cortical areas supplied by the MCA.

5. **The answer is B.** The patient has unilateral neglect, which occurs with a nondominant parietal lobe lesion.

6. **The answer is B.** Lower face weakness on the right occurs if the adjacent primary motor cortex containing upper motor neurons is involved in the lesion.

7. **The answer is A.** REM, or paradoxical, sleep is characterized by a sudden conversion from a synchronized to a desynchronized EEG. Brainstem cholinergic neurons in the ascending arousal system show increased activity and act as "REM-on" cells to initiate REM sleep.

8. **The answer B.** A lacunar stroke involving the lenticulostriate branches of the MCA that supply the genu and posterior limb of the internal capsule on the right accounts for these symptoms.

9. **The answer is C.** The most common location for a complex partial seizure is in the temporal lobe. In this patient, it was preceded by an olfactory aura.

10. **The answer is B.** Gerstmann's syndrome results from a lesion to the inferior parietal lobule, including the angular gyrus in the left hemisphere. These patients have acalculia (an inability to perform simple arithmetic problems), finger agnosia (an inability to recognize one's fingers), and possibly alexia with agraphia.

11. G
12. A
13. P
14. H
15. C
16. C
17. H
18. D
19. E
20. Q
21. O
22. N
23. R
24. B
25. I
26. J
27. D

INDEX

Note: Page numbers followed by *t* and *f* indicate tables and figures, respectively.